AF577219

BIOTHERMODYNAMICS

MONOGRAPHS IN
MOLECULAR BIOPHYSICS AND BIOCHEMISTRY

Edited by
H. Gutfreund,
Department of Biochemistry,
University of Bristol

The Physical Behaviour of Macromolecules with Biological Functions
S. P. Spragg

Intracellular Calcium: Its Universal Role as Regulator
A. K. Campbell

Biothermodynamics: The Study of Biochemical Processes at Equilibrium
J. T. Edsall and H. Gutfreund

BIOTHERMODYNAMICS

The Study of Biochemical Processes at Equilibrium

J. T. Edsall
Department of Biochemistry and Molecular Biology
Harvard University

and

H. Gutfreund
Department of Biochemistry
Bristol University

JOHN WILEY & SONS

Chichester · New York · Brisbane · Toronto · Singapore

Reprinted with corrections August 1984

Library of Congress Cataloging in Publication Data:

Edsall, John Tileston, 1902–
Biothermodynamics: the study of biochemical processes at equilibrium.

(Monographs in molecular biophysics and biochemistry)
Includes index.
1. Thermodynamics. 2. Biological chemistry.
I. Gutfreund, H. II. Title. III. Series. [DNLM:
1. Biochemistry. 2. Thermodynamics. QU 34 E24b]
QP517.T48E37 1983 574.19′283 82-15971

ISBN 0 471 10257 1

British Library Cataloguing in Publication Data:

Edsall, J. T.
Biothermodynamics.—(Monographs in molecular biophysics and biochemistry)
1. Biological chemistry 2. Thermodynamics
I. Title II. Gutfreund, H.
III. Series
574.19′283 QH345

ISBN 0 471 10257 1

Typeset by Activity, Salisbury, Wilts.,
and printed by Pitman Press, Bath, Avon.

To
Margaret Dunham Edsall
and Mary Gutfreund

Contents

Preface

The principles of thermodynamics, and their experimental applications, have been fundamental in biochemistry, physiology, and biophysics for many years, and their importance is growing. Our aim in this short volume is to present the basic principles of thermodynamics, and to illustrate them by discussion of numerous characteristic examples of studies related to biochemical and biophysical problems. The use of certain mathematical techniques, especially partial differentiation, is of course essential for the treatment. We have endeavored to develop the mathemathical treatment in a manner that is sufficiently rigorous, but with the endeavor to provide clear indications of intermediate steps in derivations. The omission of such steps, as we are aware, can cause much trouble to many students.

Chapter 1 presents a brief historical introduction and Chapter 2 is intended as a descriptive survey of the principles of equilibrium thermodynamics, with a brief indication of the possible approaches when non-equilibrium systems are being considered. This treatment should give the student a basic idea of the scope of the subject and of its applications to biochemical problems. A significant change in style will be noted when the more rigorous treatment of equilibria in terms of chemical potentials is presented in Chapter 3.

Biochemical systems contain many components, and the magnitude of the interactions between them is commonly far greater than in most of the systems studied in pure chemistry. Thus the thermodynamic linkage relations between components, as developed particularly by Jeffries Wyman, are of major importance. We introduce them briefly in Chapter 3, and extend the treatment in Chapter 5, as an essential part of our discussion of ligand binding, and of cooperative and anticooperative interactions. We have presented much of our discussion in this chapter in terms of hemoglobin and its interactions, which furnish a rich variety of illustrations of fundamental principles.

In Chapter 3 we have emphasized the thermodynamic properties of water and aqueous solutions, since water is of such central importance in biochemistry. This chapter also emphasizes the thermodynamics of hydrophobic interactions, including the properties of gases in aqueous solution, which are fundamental for many biochemical processes.

We are concerned with equilibrium thermodynamics, and have only referred briefly to thermodynamics of irreversible processes. However, especially in Chapter 4, we have discussed the thermodynamics of some enzyme-catalyzed reactions, in cases where an intermediate is effectively in equilibrium with the reactants during the course of the reaction. This also includes systems of biological importance where the enzyme is present at a concentration comparable to that of the substrates that are undergoing reaction.

Calorimetric studies of biochemical systems are rapidly increasing in importance, and readily available microcalorimetric equipment is now highly accurate and convenient. The use of calorimetry in the study of biochemical processes will surely grow, and will prove an increasingly powerful method of obtaining answers to many biochemical problems. Our final chapter is devoted to calorimetry.

We have aimed to develop concepts and illustrate their use by discussion of experimental results, but have made no attempt to describe experimental techniques or procedures.

The idea of such a book as this originated about 1975, when both of us were members of the Interunion Commission on Biothermodynamics, which included representatives of the International Unions of Chemistry, Biochemistry and Biophysics. We received helpful advice and encouragement from our fellow members on the Commission, especially from Ingemar Wadso', William P. Jencks, Rodney L. Biltonen, and the late George T. Armstrong. This book, however, is an independent enterprise, and is not in any way the responsibility of the Commission. We also thank Hilary Muirhead for a critical reading of Chapter 5. This book has undergone a slow incubation, with many exchanges and revisions of drafts of chapters, and with meetings twice a year in Bristol or in Cambridge, Massachusetts. We have aimed for continuity and coherence in presenting the subject matter, but have not attempted to obscure the stylistic individuality of the two authors.

We hope that the book will promote increased use of thermodynamic reasoning among biochemists and biophysicists, and will enable them to realize the power of thermodynamics in providing a deeper insight into many biochemical problems.

The authors wish to thank their respective secretaries Judith Guidotti and Audrey Harvey for their patience in repeated typing of revised versions of some of the sections. Mary Gutfreund gave considerable help in the preparation of the index and in proof reading.

One of us (J.T.E.) wishes to thank the U.S. National Science Foundation for a grant (SOC 7912543) which has provided valuable help for his work.

J.T. Edsall
H. Gutfreund

December 1982

Acknowledgements

The author wishes to acknowledge the cooperation of the following publishers for granting permission for the reproduction of diagrams from their publications:

Academic Press, Inc., New York
Figure 5.16 (Rossi Fanelli et al, 1964, Adv. Protein Chem., 1973)
Figure 6.5 (Privalov, 1979, Adv. Protein Chem., *33*, 167)

Acta Chemica Scandinavica, Stockholm
Figure 3.3 and 3.4 (Hvidt, 1978, Acta Chem. Scand. *A32*, 675)

American Chemical Society, Washington, D.C.
Figure 3.1 (Angell et al, 1973, J. Phys. Chem., *77*, 3092)
Figure 4.4 (Phillips et al, 1963, Biochemistry, *2*, 501)
Figure 5.4 (Scatchard et al, 1950, J. Amer. Chem. Soc., *72*, 535)
Figure 5.11 (Nozaki et al, 1957, J. Amer. Chem. Soc., *79*. 2123)
Figure 6.1 (Sturtevant, 1955, J. Amer. Chem. Soc., *77*, 255)

American Society of Biological Chemists, Bethesda, MD
Figure 3.9 (Green, 1931, J. biol. Chem., *93*, 495)
Figure 3.10 (Green, 1931, J. biol. Chem., *93*, 517)
Figure 5.6 (Bernhard, 1956, J. biol. Chem., *218*, 961)
Figure 5.12 (German & Wyman, 1937, J. biol. Chem., *117*, 533)
Figure 6.3 (Kodama et al, 1977, J. biol. Chem., *252,* 8085)

Elsevier Biomedical Press, Amsterdam
Figure 4.5 (Rosing & Slater, 1972, Biochim. Biophys. Acta., *267*, 275)

Verlag Chemie GMBH, Weinheim
Figure 3.5 (Brun & Hvidt, 1977, Ber. Bunsenges. Phys. Chem., *81*, 930)

CHAPTER 1

Historical introduction and the aims of biothermodynamics

Thermodynamics is a subject of interest to practitioners of many disciplines. Astronomers, physicists, engineers, geologists, chemists, and biologists all have this one common grammar to their scientific language: the laws of energy utilization and interconversion. However, from a few basic laws and relationships the subject branches out into quite diverse applications. Nonetheless, if the basic principles are well understood the common problems of quite different subjects can often be seen through thermodynamics. In this way this interdisciplinary subject serves the laudable purpose of helping scientists to have a common interest.

The early history of thermodynamics is, of course, closely connected to the development of calorimetry. Heat measurements long preceded optical measurements on the progress of chemical and biological processes. Quantitative records of changes in the transmission of light at a specified wavelength, the most common analytical tool in a modern biochemical laboratory, are a very recent development compared with heat measurements. There was notable progress in the eighteenth century in the development of thermometry and measurements of heat capacities and other physical constants of well defined substances. The history of the development of calorimetry in chemistry is well described by Armstrong (1964), who lists the interpretive achievements side by side with those in the construction of ingenious instruments.

The intake of energy in the form of food, its storage in the organism, and its release in metabolic activity are fundamental to biochemistry and biophysics. Among the vast array of biochemical reactions, many proceed spontaneously if catalysed by suitable enzymes. Others of central importance, such as the biosynthesis of nucleic acids, proteins, and many other substances, cannot proceed spontaneously, but must be coupled to other spontaneous reactions that can provide the necessary free energy to make the synthetic process go. Living organisms transform chemical into mechanical energy, as in the contraction of muscles or the beating of cilia, and thermodynamic relations are vital to the understanding of such processes. Processes that are potentially spontaneous often do not proceed at any appreciable rate; in biochemistry this means generally that a specific enzyme must be present to catalyse the process. Such restrictions are of course necessary for the economy of the organism, to prevent wasteful release of energy when it is not needed.

Equilibrium thermodynamics, which is what we shall discuss in this book, cannot predict how fast a process will go, but can tell whether it can go at all, and how far it can go before the process comes to equilibrium.

Much of our information in such studies comes from the direct measurement of chemical equilibria, if necessary in the presence of a suitable enzyme or other catalyst to hasten the attainment of equilibrium. We also want to know how such equilibria are shifted by changes of temperature, pressure, and other variables, or by changes in the solvent medium in which the process takes place. Accurate calorimetric studies of the quantity of heat absorbed or released in the various processes are also fundamental for our knowledge of chemical equilibria. Indeed such thermal measurements, including determination of heat capacities of the substances involved, carried down to temperatures approaching absolute zero, can permit the calculation of chemical equilibria even when they have not been actually measured.

Thermodynamics, as it has developed in the nineteenth and twentieth centuries, grew out of earlier developments that occurred mainly in the eighteenth century. On the other hand, there was the early development of calorimetry, by such men as Joseph Black and Lavoisier. Indeed Lavoisier and Laplace inaugurated the study of biochemical calorimetry in 1784 by measuring the heat evolved by a small mammal when metabolizing carbohydrate, and showing that it was the same, within experimental error, as the heat of combustion of the same substance in a calorimeter.

However, the major practical development that later led to the science of thermodynamics was the invention of heat engines that transformed the heat of steam, under pressure, into mechanical work. The heat engines of Thomas Savery, Denis Papin, and Thomas Newcomen, in the early eighteenth century, were used for such purposes as pumping water out of mines. By modern standards they were grossly inefficient; nevertheless they found use. The important improvements made by James Watt, from 1763 on, made these devices much more efficient, and brought them into wider use; but all these developments occurred on a purely empirical basis, without a theoretical foundation (see for instance Cardwell, 1971).

The first major theoretical advance came in 1824 from the French engineer Sadi Carnot, who in his great memoir on the motive power of heat (Carnot, 1824) formulated the ideal conditions for the operation of an engine that transformed heat into mechanical work and calculated the maximum possible efficiency that such an engine might attain. He pictured the engine as receiving heat from a boiler at temperature T_1 and discharging it to a condenser at a lower temperature T_2, the process involving two cycles of expansion and compression in a cylinder in which the working fluid (usually, but not necessarily, steam) drove the motion of a piston. We need not discuss the details of the Carnot cycle here; they are given in most text-books of thermodynamics. Carnot's general conclusion, however, was of immense general importance. The maximum possible efficiency of the engine – that is, the ratio of the work produced to the input of heat – depended on the temperature difference ($T_1 - T_2$) between the boiler and the condenser, and only a fraction of the thermal energy could be converted into work. However, the reverse process – the complete dissipation of mechanical energy into heat – was all too easy. Thus Carnot's work pointed to a fundamental irreversibility in natural processes; and in this respect he anticipated what later came to be known as the second law of thermodynamics. When he wrote his great memoir Carnot, like many of his contemporaries, still thought of heat as a caloric fluid which could pass from one body to another; the concept of the conservation of energy and the absolute scale of temperature had not yet been established, so that Carnot's analysis was still incomplete. There is evidence from later notes, unpublished until long after his death, that he was working toward the concept of conservation of energy; but he died in a cholera epidemic in 1832, at the age of 36. A recent article by Wilson (1981) gives a valuable portrayal of Carnot and his work and the book by Cardwell (1971) covers the whole development from the early steam engines to the enunciation of the first and second laws of thermodynamics.

The conception of heat as a form of energy, and the quantitative conversion of one form of energy into another, became definitely established in the decade after 1840. The idea of conservation of energy was put forward by a German physician, Julius Robert Mayer, a man fertile in ideas and interpretations. It was James Prescott Joule, however, whose careful experimental studies of the conversion of electrical and mechanical energy into heat established quantitatively the interrelation between different forms of energy (see particularly Joule, 1849). It was the great memoir on conservation of energy by Hermann von Helmholtz (1847), however, which provided a systematic formulation of the concept that was decisive for the general acceptance of the first law of thermodynamics. (It is interesting to note that Poggendorf, editor of the *Annalen*, rejected the paper when Helmholtz submitted it to him.) Helmholtz was almost certainly the most productive biophysicist of all time (Konigsberger, 1965). He demonstrated, while still a young army surgeon, that metabolic energy was quantitatively transformed into work and heat in muscle. He made immense contributions to physiological optics and auditory physiology, and was the first to measure the speed of conduction in nerve, in addition to his work on thermodynamics. He ended his career as Professor of Physics in Berlin, having earlier held chairs of Anatomy and Physiology at several universities.

Since that time studies of the mechanism of energy transduction have become a central theme in biophysics. The energy which becomes available during the metabolic breakdown is utilized for such diverse processes as the synthesis of nucleic acids (information storage) and proteins (functional and structural) as well as for the transmission of information in nerve conduction, the synthesis and transmission of messengers, and many mechanical processes related to muscle contraction. The compounds which undergo this metabolic breakdown for the provision of energy are, in turn, synthesized in the photosynthetic processes which utilize the energy of the electromagnetic radiation received from the sun. An interesting discussion of the efficiency of the food chain, from an ecological point of view, is contained in a recent popular book (Colinvaux, 1980).

The fundamental importance of Carnot's work took some time to be appreciated. Its recognition came from about 1850 on, thanks largely to William Thomson (Lord Kelvin) in Great Britain and Rudolf Clausius in Germany. It was Kelvin who first established, on a rigorous basis, the thermodynamic (absolute) scale of temperature – a concept that indeed had been under discussion by various authors from about the beginning of the nineteenth century. Making use of this concept, Kelvin was able to formulate more exactly Carnot's equations for the maximum efficiency of an ideal heat engine, working between the temperature (T_1) of the boiler and that (T_2) of the condenser. The maximum ratio of the work done to the heat input was given by the simple expression

$$\text{work/heat energy} = (T_1 - T_2)/T_1.$$

Thus the possible yield of work depended on the temperature difference and was always less than the input of heat energy, except in the impossible case of a condenser held at a temperature of absolute zero. It is to Kelvin, and especially to Clausius, that we owe the development of the concept of entropy (from the Greek word τροπη: transformation), which is central to the second law of thermodynamics. Entropy, unlike energy, is not a quantity that remains constant, independent of time. Instead, for any closed system that cannot exchange matter or energy with its surroundings, any spontaneous process that may occur must involve an *increase* of entropy. Entropy is a measure of the irreversibility of natural processes. We deal more fully with this powerful and subtle concept in Chapter 2, and shall make constant use of it in the rest of this book.

Snow (1959) states, in his provocative essay on 'the two cultures' that it should be just as much part of a prerequisite for an educated man to be able to talk about the second law of thermodynamics as it is to have read the plays of Shakespeare. There are many very deep analyses of the second law still debated in the current literature, but then the same is true in expert interpretations of Shakespeare.

It is a characteristic feature of biological processes that energy interconversion occurs under conditions of negligible or very small temperature gradients – they are essentially isothermal throughout. Needham (1971) describes how, towards the end of the nineteenth century, it became clear that the maximum temperature gradients which could be encountered in muscle could not result in the efficiency realized in terms of a heat engine. Fick's (1893) comments on the origin of muscle power are of interest in this connection. The mechanism of muscle contraction must depend on direct coupling between chemical and mechanical energy. While the complete molecular details of this process are not yet understood, the combined results of investigations into the structure of muscle fibres, as well as their physiological and molecular responses to stimulation by a nerve impulse, have provided plausible models for the mechanism of energy transduction in this system (see for instance Keynes and Aidley, 1981). Several different aspects of this important problem in biothermodynamics will be discussed in various sections throughout this volume.

The man who laid the foundations of chemical thermodynamics in the late nineteenth century was J. Willard Gibbs of Yale University (see Gibbs, 1876, 1931). Gibbs considered the relations involved in the establishment of equilibrium, for systems composed of any number of chemical components and phases, as functions of temperature, pressure, and chemical composition. He also showed how the analysis could be extended to include other forms of energy, such as surface energy or electrical energy, when these might become important. He also defined new thermodynamic functions, such as the enthalpy (heat content) and the functions now known as the Helmholtz energy and the Gibbs energy (the latter has commonly been called the free energy by many chemists). These functions are of the utmost importance in thermodynamics today, and we shall make constant use in this book of the enthalpy and the Gibbs energy.

Gibbs wrote in a highly condensed and mathematical style, and few chemists of his generation could understand or make use of the powerful methods he developed. He did, however, greatly influence the group of physical chemists led by J. H. van't Hoff, Svante Arrhenius, and Wilhelm Ostwald, who from about 1885 on developed the physical chemistry of solutions and the concept of electrolytic dissociation. It was primarily G. N. Lewis and his school in California, in the early twentieth century, who showed how to apply the methods of Gibbs to experimental data in the study of chemistry, and the book on thermodynamics by Lewis and Randall (1923) had a profound influence on a whole generation of chemists and biochemists.

Another great landmark in the development of the subject was the 'heat theorem' of Walther Nernst (1906) which became the basis of the third law of thermodynamics. This made possible, at least for many systems, the calculation of chemical equilibria from purely thermal measurements: determinations of heat capacities, and of the heat evolved or absorbed in phase transitions. This we discuss further in Chapter 2.

As more and more specialized interests have developed in the study of energy relationships in biology, a division has occurred between those who are primarily interested in the storage of oxidative and photosynthetic energy (the bioenergeticists) and those who are interested in the utilization of such stored energy for biosyntheses, muscle contraction, transport, and the transmission and transduction of information. The subject of bioenergetics is now largely concerned with the mechanisms of oxidative phosphorylation and photosynthesis. These are the two main processes for the production of ATP (adenosine triphosphate). Many objections have been raised to the terminology used in relation to ATP synthesis and utilization. The terms 'energy store' and 'high energy phosphate bonds' have been much criticized. Such semantics has contributed nothing to the progress of science. It is hoped that the definitions and explanations given in this volume will be unambiguous and a reasonable approach to the correct ones.

ATP can be called the energy store of biological systems because of its special role at the junction between its formation during oxidative and anaerobic metabolism as well as photosynthesis and its utilization by energy-requiring processes. The principal energy consumption is due to the biosynthesis of macromolecules, muscular and other movement, the transport of ions against concentration gradients, and a great variety of control processes. This function is due to its characteristic physicochemical properties, especially of its phosphate ester bonds. These properties and their thermodynamic consequences are discussed in some detail in Chapter 4.

Edsall (1974) and other contributors to a symposium on the history of bioenergetics describe how the different aspects of this subject have developed during the last two centuries. In that collection of papers it is clearly seen that an understanding of the energy relationships went side by side with the exploration of chemical pathways.

Some other areas of biochemistry and physiology should be mentioned in which the study of energy changes has made interesting contributions. Calorimetric measurements have been carried out on biological systems at different levels of organization: the reader is referred to the surveys by Kleiber (1961) and Ihde and Janssen (1974) for the study of mammals and to Beezer (1980) for the analysis of heat measurements of whole cells. The study of energy changes during the stimulation and function of muscle and nerve fibres is not only of interest in itself, but it has also played an important part in the history of the development of calorimetry. A. V. Hill's classical investigations of heat changes in active nerve and muscle fibres spanned nearly 60 years from 1912 (see Hill, 1965). Hill and his colleagues (see Howarth, 1970) were able to measure the progress of reactions which resulted in 5 μK temperature changes during a time course of 50 ms after stimulation of nerve or muscle fibres. While modern advances in electronics have been applied by other investigators, it is of interest that the resolution obtained by Hill with thermopiles and galvanometer systems, designed in his laboratory, has not been surpassed.

Advances in the interpretation of experiments on muscle in terms of energy balance come from the combination of heat measurements with other methods of analysis. Modern techniques of observing rapid chemical and spectroscopic changes during muscle contraction, as well as accurate calorimetric data on the chemical changes which occur during contraction and recovery, are continuing to increase the insight into the thermodynamics and mechanism of muscular activity (for a review of work from many laboratories see Homsher and Keen, 1978). This is of particular interest since these studies are going on side by side with a more and more detailed understanding of the molecular events during the interaction of well defined proteins in a muscle.

In connection with the above application of heat measurements on complex systems it should be pointed out that the lack of specificity of heat measurements can be a major contribution to the discovery of new phenomena. If the measured heat production or uptake of the organized system does not correspond to the sum of the heat changes during the known individual processes which take place, then one has evidence for additional previously neglected component reactions. This has turned out to be the case in muscle contraction, where an initial excess heat production has been discovered (see review by Homsher and Keen, 1978). This approach is likely to give similar useful information on other complex systems as calorimetric studies are being more widely used to monitor the progress of reactions in organized systems and in isolation. A range of calorimeters for different purposes is continuously being developed and these calorimeters are becoming commercially available. These techniques are, however, still grossly underused, considering their potentiality.

During the period 1930–50 a considerable research effort was directed towards the energy requirements for the biosynthesis of proteins – which were at that time thought to be the only macromolecules with genetic and chemical specificity. Calorimetric and other data were collected which permitted calculations of the thermodynamic data for amino acids, peptides, and proteins (see for instance Borsook and Deasy, 1951). This is not the place to review the investigations which demonstrated during the last 25 years how – in chemical terms – the energy of hydrolysis of the phosphate ester bonds of ATP is used for the synthesis of peptide bonds. Similarly phosphate ester bond hydrolysis of several nucleotide triphosphates is coupled to the specificity and control of polypeptide chain assembly on microsomes, with the genetic pattern provided by the nucleic acid template. The replication of the nucleic acids is in turn coupled to the hydrolysis of nucleotide phosphate esters.

Conceptually it is easier to understand how calorimetry can provide information about the energy balance for a sequence or for parallel processes, as compared to other thermodynamic information, which requires a deeper understanding of the subject for conversion of experimental data into thermodynamic parameters. These other methods involve the interpretation of equilibrium measurements on a large variety of different processes. A list of phenomena of biological interest which can be observed at equilibrium includes osmotic pressure, solubility, electrode potentials, oxidation–reduction potentials, ionization involving protons and other ions, binding equilibria of substrates to macromolecules, protein–protein interaction, and, last but not least, chemical equilibria. Many research workers interested in the physicochemical aspects of biochemistry and physiology have not only stimulated but have made major contributions to our present understanding of the behaviour of electrolytes, the properties of ionizing groups, and the energies of other non-covalent interactions in aqueous systems (see Cohn and Edsall, 1943; Edsall and Wyman, 1958; Tanford, 1961). The importance of the physical chemistry of aqueous solutions has been recognized in most text-books on general physiology. The contributions of biochemists and their interest in oxidation–reduction potentials and other equilibria for the understanding of energy relationships were surveyed in a symposium held at the New York

Academy of Sciences some 30 years ago and has been maintained since, in connection with the coupling of oxidative processes to the phosphorylation of nucleotide phosphate. In the continuing endeavour to understand the chemical and physical processes connected with this important problem in biochemistry (see Boyer *et al.*, 1977) a careful watch on the restrictions set by the laws of thermodynamics and their consequences acts as a continuous guide towards an acceptable solution.

The complexities of even the simplest organisms lead both to the great intellectual interest and to the many misleading statements made about the thermodynamics of biological systems. In thermodynamics, as in any other branch of science, precise definitions of the terms used are essential. Having provided some background to the scope of the subject it is now essential to present some of the basic definitions of concepts involved in the interconversion of different forms of energy as well as the relations between temperature, pressure force, energy, work, and power with their units and dimensions. The second chapter will conclude the elementary survey of thermodynamics so that the rest of the volume can be devoted to building detail and applications on a firm foundation. In the later chapters of this volume we shall cover the theoretical background and provide numerical examples to help in the design of experiments and the interpretation of data obtained from them without dealing with the nuts and bolts of the experimental procedures and equipment. We shall concern ourselves with the thermodynamic analysis of the behaviour of solutions of defined mixtures of molecules and their reactions in homogeneous systems. These solutions contain the components of physiologically active systems and a part, and at times the whole, of the organization necessary for functions such as catalysis, control, muscle contraction, and the transduction of information. However, the emphasis is on the characterization of the components of biological systems.

Equilibrium thermodynamics and calorimetry provide the data for enthalpies, entropies, and Gibbs functions of the formation and conformational equilibria of macromolecules and of the catalytic and ligand binding processes in which they are involved. These processes are basic to the understanding of the behaviour of biological systems; even so they are not in themselves complete. The fully organized systems with selectively permeable membranes require additional thermodynamic concepts. The scope of thermodynamics is so wide that no one volume can treat all aspects of it. The application of thermodynamics to organized systems in a functional steady state, which would describe an intact biological system, clearly requires separate coverage from the ground rules given here (see Section 2.7). Similarly the connection between thermodynamics and rate processes would be a natural continuation of the subjects covered in this volume. Both these topics will get a brief mention in Chapters 2 and 4 so that their relation to equilibrium thermodynamics is defined.

Throughout this text it is often necessary to use a term or mention a concept which has not been defined at that stage. In such cases reference will be made to the section in which a more detailed explanation can be found, unless the particular topic is beyond our scope. This is especially relevant to the next chapter, which deals with the fundamental concepts of thermodynamics. That chapter can be read as an elementary introduction to the subject, while Chapter 3 represents a detailed analytical treatment of thermodynamic functions. Chapters 4, 5, and 6, which comprise the second half of the book, are concerned with the interpretation of phenomena most widely studied in biochemical investigations of thermodynamic parameters.

CHAPTER 2

Fundamental concepts: phenomenological

2.1 Systems and efficiency

In thermodynamic discussions the word 'system' should be used with care and in a well defined manner. A system is any specified part of the universe. It is geometrically defined by a boundary and some important aspects of its properties depend on the characteristics of its boundary.

An isolated system has a boundary which is impermeable to heat and matter. Any process occurring in such a system is defined as adiabatic. In the strictest sense it would be impossible to make measurements on isolated systems without any communication with the outside world. However, for practical purposes one can approach 'isolated conditions' sufficiently closely, with corrections for deviations, in, for example, a Dewar flask.

A closed system has boundaries permeable to all forms of radiation including heat, but impermeable to matter. This can be maintained as an isothermal system, which is of particular relevance to the study of reactions in solutions and of the thermodynamic properties of components of biological processes.

An open system is a geometrically defined volume which can exchange both heat and matter with its surroundings. As stated in Chapter 1, this is the kind of system generally found in biological organisms, although there are usually some specific restrictions on the permeability to selected solutes. As will be seen later, in addition to definitions of the boundary, careful statements about constancy of temperature, pressure, volume, etc., are also required.

When one considers the efficiency of a process, the strictly thermodynamic definition does not take into account the requirements of the organized biological systems as a whole. In the thermodynamic sense one talks about efficiency as defined for a steam engine (see Kauzmann, 1967, pp. 121–127) or the heat loss during battery charging. Often heat is referred to as waste – for instance, that due to friction in a pulley. In a warm blooded animal the production of heat serves a purpose and under many conditions it is a useful by-product of metabolic processes such as the hydrolysis of ATP during ion transport. If the requirement for heat is greater than that provided by routine functions of the organism, the well known process of shivering utilizes metabolic energy and results in production of the heat required to keep the temperature of the body near 37 °C, in the case of humans. Some tissues have special mechanisms for heat production through uncoupled oxidation. Other aspects of temperature control are discussed in Section 2.6 in connection with the functions of water as a thermostat and in Section 4.8 in connection with the energy dissipation during metabolic control.

In considering such questions as the thermodynamic efficiency of living organisms, we must bear in mind their special needs and functions, which are often profoundly different from those of man-made machines. For instance, as we shall see later (Chapter 4), the minimum amount of energy input required to bring about the linkage between two amino acids to form a peptide bond has been accurately measured in several instances. However, the biosynthesis of a peptide chain in a protein involves an elaborate series of steps that altogether require about 10–20 times as much energy, per peptide bond formed, as this basic minimum. It would seem at first sight that nature has developed a method of protein synthesis that is wildly extravagant in its use of energy. However, in return for this expenditure the organism gets one of its basic requirements. The amino acids are linked together in a definite order, prescribed by the genetic code and specified by the sequence of bases in the messenger RNA that codes for the protein. The organism can afford to pay a high price in energy consumption in order to achieve specificity. Without the specificity that enables a protein to perform a definite function, for instance as a specific enzyme, there would be no point in stringing together a random sequence of amino acids. In terms of the thermodynamic quantities that we discuss later in this chapter, we can say that the organism expends energy in order to reduce the entropy of the specific product. Low entropy corresponds to increased order, high entropy to increased disorder (cf. Edsall, 1974, pp. 106–108).

2.2 Physical quantities: units and dimensions. Interconversion of different forms of energy

It is important to distinguish between physical quantities, such as mass, length, time, temperature, and energy, and the units in which those quantities are expressed. We cannot give a numerical value corresponding to a length, for instance, until we have chosen a suitable unit of length. Table 2.1 lists a number of important physical quantities, and the units used in the Système Internationale (SI) to express numerical values of these quantities. It is an important convention, which we follow in Table 2.1, to denote physical quantities by sloping (italic) letters and units by upright (Roman) letters. Thus m is the symbol for the quantity mass (the unit being the kilogram) whereas m is the symbol for the unit of length, the metre.

Table 2.1 also lists the physical dimensions of many of these quantities. To do this we must choose certain quantities as primary, and express others in terms of these. It is generally most convenient to take mass (m), length (l), time (t), and temperature (T) as primary quantities, and we can then express many other quantities in terms of these. We are under no compulsion to choose

Table 2.1 Physical quantities, symbols and dimensions: SI units and their symbols

(a) Base quantities

Quantities	Symbol	Name of unit	Symbol for unit
Length	l	metre	m
Mass	m	kilogram	kg
Time	t	second	s
Electrical current	I	ampere	A
Temperature	T	kelvin	K
Amount of substance	n	mole	mol
Luminous intensity	I_v	candela	cd

(b) Other derived quantities

Quantities	Symbol and (dimensions)	Units	Symbol and dimensions for units
Force	$F\ (mlt^{-2})$	newton	N (kg m s^{-2})
Pressure	$p\ (ml^{-1}t^{-2})$	pascal	Pa (N m^{-2})
Energy	$U\ (ml^2t^{-2})$	joule	J
Power	$P(ml^2t^{-3})$	watt	W (J s^{-1})

Electrical charge	$Q\ (It)$	coulomb	C (A s)
Electrical potential difference	$\mathbf{E}\ (UI^{-1}t^{-1})$	volt	V ($J\ A^{-1}\ s^{-1}$)
Electrical resistance	$R\ (\mathbf{E}I^{-1})$	ohm	Ω ($V\ A^{-1}$)
Frequency	$\nu\ (t^{-1})$	hertz	Hz (s^{-1})
Area	$A\ (l^2)$	square metre	(m^2)
Volume	$V\ (l^3)$	cubic metre	(m^3)
Density	$\rho\ (ml^{-3})$	kilogram per cubic metre	($kg\ m^{-3}$)

(c) Prefixes

Fraction	Prefix	Symbol	Multiple	Prefix	Symbol
10^{-1}	deci	d	10	deca	da
10^{-2}	centi	c	10^2	hecto	h
10^{-3}	milli	m	10^3	kilo	k
10^{-6}	micro	μ	10^6	mega	M
10^{-9}	nano	n	10^9	giga	G
10^{-12}	pico	p	10^{12}	tera	T
10^{-15}	femto	f	10^{15}	peta	P
10^{-18}	atto	a	10^{18}	exa	E

Table 2.2 Useful numbers and conversion factors

Avogadro number $N = 6.022 \times 10^{23}$ mol^{-1}
Gas constant $R = 8.314$ J K^{-1} mol^{-1} (= 0.0821 l atm K^{-1} mol^{-1})
Boltzmann's constant $k = R/N = 1.3807 \times 10^{-23}$ J K^{-1}
Charge of a proton $e = 1.602 \times 10^{-19}$ s A
One faraday (i.e. 1 F) = $N \times e = 9.648 \times 10^{4}$ s A mol^{-1} = 9.648×10^{4} C mol^{-1}
One coulomb (i.e. 1 C) = 6.28×10^{18} electronic charges = 2.998×10^{9} e.s.u. (electrostatic units)
One debye (i.e. 1 D) = 10^{-18} e.s.u. cm
One electron volt (i.e. 1 eV) = 96.47 kJ mol^{-1}
One calorie = 4.184 J
One standard atmosphere (at 298.25 K) = 1.013×10^{5} Pa (760 cm Hg)
Planck's constant $h = 6.626 \times 10^{-34}$ J s
One dyne = 10^{-5} N
One erg = 10^{-7} J

this set of primary quantities, however; on occasion, for instance, it may be convenient to choose energy (U) as a primary quantity, rather than treating it as the product of a mass multiplied by the square of a velocity, as we have done in Table 2.1. We must of course be consistent in our choice of dimensions in any particular set of calculations (Bridgman, 1931).

Analysis of physical quantities in terms of dimensions is often valuable in making sure that all the terms in equations are consistent. All terms that are added in an equation, or set equal to others, must have the same dimensions. Thus in the equation for a perfect gas, $pV = nRT$, where n is the number of moles of gas, p has the dimensions of force per unit area ($mlt^{-2}/l^2 = ml^{-1}t^{-2}$) and volume is of dimension l^3. Hence pV is of dimension ml^2t^{-2} and has the dimensions of energy. Here n is a pure number, and therefore is dimensionless. Hence the gas constant R must have the dimensions of energy divided by temperature, so that RT will have the dimensions of energy. We shall usually express R in joules per kelvin per mole (J K^{-1} mol^{-1}).

Often the base units in Table 2.1 are too large or too small to give convenient numerical values for the quantities under study. Fractions or multiples of these units may be much more convenient. The system of prefixes listed in the bottom section of the table provides for this. For instance a large electrical power plant may produce energy at a rate of 10^9 J s^{-1} or 10^9 W. This is 1 gigawatt (1 GW) and if we are talking about power of this magnitude it is convenient to give values in gigawatts. If we are concerned with the wavelengths of visible light, they are of the order of a few hundred nanometres (1 nm = 10^{-9} m). Interatomic distances are of the order of 0.1 nm, or 100 pm; for instance the C—C distance in ethane is 0.154 nm or 154 pm. (This is of course the same as 1.54 Å, where the Ångstrom unit (Å) equals 10^{-8} cm. This unit is compatible with the SI system but is not generally considered a part of it.) The SI unit of pressure, the pascal (Pa) is a small unit.

One standard atmosphere (1 atm) is equal to 101 325 Pa or 101.325 kPa. (Another pressure unit, the bar, equals 100 kPa exactly; but the bar, like the atmosphere, is not a SI unit.)

Dimensional analysis is useful for converting numerical values from one set of units to another. For instance the density of cold water is close to 1 g cm^{-3}. What is this value if expressed in kilograms per cubic metre? We know that 1 g = 10^{-3} kg, and 1 cm = 10^{-2} m. Hence 1 g cm^{-3} = 1×10^{-3} kg/$(10^{-2})^3$ m^3; rearranging, we find the density is 10^3 kg m^{-3}. The same procedure can be followed in more complicated cases (Bridgman, 1931).

In the chemical and biochemical literature, the calorie is still widely used as a unit of energy. One calorie was originally defined as the amount of heat energy required to raise the temperature of 1 g of water by 1 °C (= 1 K). However, the actual amount of energy required varies slightly, but significantly, with the initial temperature of the water. To obviate confusion the defined calorie is now set equal to exactly 4.184 J. The joule is gradually displacing the calorie as the unit of choice for reporting energy data, but workers must be familiar with both units, and with the relation between them.

A much larger list of physical quantities and units is to be found in the IUPAC *Manual of Symbols and Terminology* (revised edition 1975). For a still more recent edition of this report, see Whiffen (1979).

It has become common practice to label graphs and tables in such a way that the results are listed as pure numbers. This is done by using in the heading or legend the symbol for the quantity divided by the symbols for the units. For instance, if a column gives the values for the heat evolved per mole of substance reacted, the column should be headed

$$q/\text{kJ mol}^{-1}.$$

Similarly, some books and journals now use pure numbers in the text and instead of writing for 40 kJ evolved during a reaction

$$q = -40 \text{ kJ mol}^{-1}$$

one can write

$$q/\text{kJ mol}^{-1} = -40.$$

It will become clear how useful the concepts of units and dimensions are when we consider the problems connected with the different forms of energy. The greatest need for a thorough understanding of thermodynamics in biology comes from the desire to explain energy transformations.

Biological systems are particularly versatile in their ability to interconvert mechanical, electrical, chemical, light, and even magnetic energy. In their sensory and control functions they also have remarkably sensitive measuring devices. From some later comments (Section 2.4) it will be seen that there is also a relation between energy and information storage and transmission.

When we discuss conservation of energy we must remember that it is the sum of all forms of energy (mechanical, thermal, electrical, etc.) that is conserved.

Energy and work are the quantities which can be used to present a unified picture of different forces and their consequences. The definition of work is given by

$$W = \int_{r_1}^{r_2} F\mathrm{d}r, \tag{2.1}$$

where $r_2 - r_1 = l$, the distance of displacement of a particle on which the force F acts. As can be seen in Table 2.1, the dimensions of force are derived from acceleration (lt^{-2}) of unit mass (m). The resulting dimensions for work ($l^2t^{-2}m$) can also be derived for energy; for example, for the particular case of kinetic energy

$$E = \tfrac{1}{2}\, mv^2 \tag{2.2}$$

from the dimensions for velocity (lt^{-1}).

Pressure ($ml^{-1}t^{-2}$) is the force per area and the relation between pressure and energy is described in terms of dimensions by

$$\text{pressure } (ml^{-1}t^{-2}) \times \text{volume } (l^3) = \text{energy } (ml^2t^{-2}),$$

which leads to the gas equation in the form

$$pV/T = R, \tag{2.3}$$

where V is the volume of 1 mol of gas and R (the gas constant) has the units joules per kelvin per mole (J K^{-1} mol^{-1}).

Another physical quantity which comes into consideration as regards the energy balance of many biological processes is the viscosity (η) of fluids. In viscous flow layers of liquid or gas slide past each other, in contrast to turbulent flow, which is irregular and more complex to analyse. A force is required to move the layers in viscous flow. This force is proportional to the area (A) and to the relative velocities (Δv) of the layers and inversely proportional to the distance (Δl) between the two outside layers:

$$F = \eta A \frac{\Delta v}{\Delta l}, \qquad \eta = \frac{F/A}{\Delta v/\Delta l},$$

where the viscosity (η) is the proportionality constant. The dimensions of viscosity are

$$\eta = \frac{mlt^{-2}/l^2}{lt^{-1}/l} = ml^{-1}t^{-1}.$$

The dimensional relation between energy and viscosity,

$$\text{energy } (ml^2t^{-2}) = \text{viscosity } (ml^{-1}t^{-1}) \times l^3t^{-1},$$

can be explained in terms of

$$\text{viscosity} = \frac{\text{energy required to maintain flow}}{\text{volume of liquid/time of flow}} .$$

The relation between pressure and viscosity is given by the practical unit (poise) for viscosity: 1 poise = 0.1 Pa s.

A discussion of power can be used to lead to the consideration of units of electrical quantities. Power is work per unit time and has the dimensions ml^2t^{-3}; this can be called the rate of energy production or rate of work performance with the units joules per second (J s^{-1}), which is equal to the practical unit known as the watt (W). The coulomb (C) is important in the definition of the fourth unit of the SI system, the ampere (A), the unit of electric current,

$$1 \text{ A} = 1 \text{ C s}^{-1},$$

and the electric potential (V),

$$1 \text{ V} = 1 \text{ J C}^{-1}.$$

Dimensional analysis shows that

$$1 \text{ W (J s}^{-1}) = 1 \text{ A (C s}^{-1}) \times 1 \text{ V (J C}^{-1});$$

i.e. one watt corresponds to a current of one ampere at a potential of one volt. If electrical power is completely converted into heat energy, one joule will raise the temperature of one gram of water (at 25 °C) by 0.239 K. This relation is useful for the electrical calibration of calorimetric equipment. The amount of power required to raise the temperature of a specified quantity of a substance by 1 K s^{-1} is dependent upon the specific heat capacity of the substance. This topic is discussed in some detail in Chapter 6. It is of interest to note in this connection that the heating effect of an electric wire can be calculated directly from its resistance and the applied voltage. The resistance in ohms (Ω) can be expressed as volts per ampere. If a constant voltage is applied to the heating wire and the current is measured, the resistance is given in ohms = volts per ampere. Since power (in watts) has the dimensions of joules per second, it is apparent from dimensional analysis that $1 \text{ W} = 1\ \Omega \times (1 \text{ A})^2$. Since the resistance is proportional to the length of the conductor and inversely proportional to its cross-sectional area, it is expressed in ohm metres.

The mechanical equivalent of heat determined by Joule (1849), and named after him, is derived as follows. One joule of energy will move an object against a force of 1 newton over 1 metre. The gravitational acceleration at sea level $\boldsymbol{g} = 9.81$ kg m s^{-2}, and the force on 1 kg is 9.81 kg m s^{-2} (9.81 newton). Hence the energy of lifting 1 kg through 1 m against this force is 9.81 J. If we use the same amount of energy to heat water, we would get the following results. We have just seen that 1 J s^{-1} will heat 1 g of water by 0.239 K; therefore, 9.81 J will raise the temperature by about 2.34 K. To conform with SI nomenclature, we should consider 1 kg of water and this would require 1 kJ for a temperature rise of 0.239 K. It should be apparent that 1 kcal (4.184 kJ) will raise the temperature of 1 kg of water by 1 K.

In a discussion of different forms of energy with a role in biological processes, one must also consider the relation between the wavelength and the energy per photon in electromagnetic radiation. The absorption of radiation by photoreceptors in photosynthetic and visual systems, as well as the emission of light by a number of organisms, require analysis in terms of thermodynamic balance during conversion to or from other forms of energy. The einstein, named after the first physicist to interpret the photoelectric effect, is the unit of which describes one mole of photons. The following calculation demonstrates the wavelength dependence of the energy per einstein. The energy of electromagnetic radiation expressed per quantum or photon is

$$E = h\nu = hc/\lambda,$$

where h is the Planck constant (6.626×10^{-34} J s), c is the speed of light in a vacuum (2.997×10^{8} m s^{-1}), and ν is the frequency corresponding to the wavelength λ. The energy corresponding to one einstein (6.03×10^{23} photons) of radiation in the visible range (for instance at $\lambda = 600$ nm) is given by

$$\frac{6.626 \times 10^{-34} \times 2.9979 \times 10^{8} \times 6.03 \times 10^{23}}{600 \times 10^{-9}} = 2.00 \times 10^{5} \text{ J}.$$

Substitution of different wavelengths in the above equation shows the inverse relationship between wavelength and energy of radiation and could, for instance, be used to calculate the heating effect of light absorbed and converted into thermal energy.

Another energy unit which is in common use in one branch of biochemistry is the electron volt (eV). It is defined as the energy per mole acquired by electrons when accelerated through a potential of 1 V:

$$1 \text{ eV} = 96.4 \text{ kJ mol}^{-1} = 23 \text{ kcal mol}^{-1}.$$

Oxidation–reduction reactions are often characterized in terms of these energy units. A more positive potential indicates a greater affinity for electrons in oxidation–reduction processes.

Some rough calculations on biological problems can illustrate the conversion of units and energy output. If a man's diet provides 2500 kcal of energy per day, how does his energy output compare with that of a 100 W electric light bulb? To a first approximation

$$2400 \text{ kcal} \simeq 10^{4} \text{ kJ},$$

$$1 \text{ day} = 86\,400 \text{ s} \sim 10^{5} \text{ s}.$$

Then the man's power output is $10^4/10^5 = 0.1$ kJ s^{-1} or 100 J s^{-1} and 100 W = 100 J s^{-1}, so the man's power output and that of the electric light bulb are roughly the same.

When one thinks about energy interconversion in biology, one of the first things which comes to mind is muscle contraction. ATP, the energy donor for muscle contraction, yields a maximum of 50 J g^{-1} of energy during its reaction with myosin. This means that a perfectly efficient muscle system could lift 1 kg to a height of 5 m utilizing about 2×10^{-3} mol of ATP. In fact muscle works at about 30% efficiency, the rest of the energy being released as heat.

Also, in a perfect physical system no work should be required to hold a weight at a given height. This would be the case in a perfectly superconducting electromagnetic system. With normal electromagnets, and with a muscle, 'maintenance' work has to be done during the holding of a weight.

2.3 The laws of thermodynamics

The laws of thermodynamics are laws of experience gained from experiments, observations, and common sense but have no theoretical basis themselves. They are the foundation upon which the theories of thermodynamics rest and they are used here to define basic principles.

The zeroth law of thermodynamics has acquired this name because it was formulated after the third law but should have come first. It defines the concept of temperature.

Temperature as a sensation tells us that some objects are hotter than others, but this provides no quantitative scale. Sensations can be deceptive. If you immerse one hand in fairly hot water, and the other in ice water, for a short time, and then plunge both hands into water at room temperature, it will feel warm to the hand that was previously cold, and cool to the hand that was hot. We need some kind of thermometer that is to serve as an objective standard. The nearest approximation to the ideal is a gas thermometer at low pressure. If we suppose that it contains 1 mol of any gas, we keep it at some fixed temperature and measure the molar volume (V) at various values of pressure (p). If we plot the product pV as a function of p, we find for any gas that pV approaches a limiting value (approximating an ideal gas) as p approaches zero. This limiting value proves to be the same for all gases. (Gases that dissociate on dilution present a special problem, but, in principle, the same analysis applies.)

To define a temperature scale and fix the size of the degree of temperature, we need at least two points, each of which corresponds to a steady and reproducible temperature. This can be achieved by first immersing our gas in a well stirred ice–water mixture, which is (by definition) zero degrees on the Celsius scale, and then immersing the gas in pure boiling water at atmospheric pressure, which has a temperature 100 degrees greater than the ice–water mixture on the Celsius scale. The limiting value of pV as $p \to 0$ is measured at both temperatures. We can then obtain

$$\lim\,(pV)_{100°} - \lim\,(pV)_{0°} = R(t_2 - t_1) = 100R.$$

R is the molar gas constant (see also p. 14, where it is expressed in units of joules per kelvin per mole (J K^{-1} mol^{-1})). It can also be expressed in units of cubic decimetres-atmospheres per kelvin per mole (dm^3 atm K^{-1}) to give the value 8.2058×10^{-2}. The value of R is the same whatever the gas used in the thermometer, it is a universal constant.

Now that we have fixed the value of R, consider again the product pV at 0 °C. We can define the absolute temperature T of the ice–water mixture by the equation

$$\lim_{p \to 0}\,(pV) = RT = 22.414 \text{ l mol}^{-1}.$$

Using the above value for R, we obtain the value 273.15 K for T at 0 °C and the value 373.15 K for T for boiling water at 1 atm pressure. Here K denotes temperature on the absolute Kelvin scale, named after Lord Kelvin who played a major role in formulating the scale. A temperature of, say, 300 kelvins is written in the SI nomenclature simply as 300 K. The degree sign is omitted, the unit is called the kelvin:

$$T = t + 273.15,$$

where t is the temperature in degrees Celsius and T the same temperature in kelvins. The size of the unit is the same on the two scales.

Many types of thermometer are in use, but they are ultimately calibrated against gas thermometers. Special problems arise in measurements near 0 K, but these are beyond the scope of the present discussion.

A statement of the zeroth law is as follows: if two systems are each in thermal equilibrium with a third system, they must be in thermal equilibrium with each other. This underlines the fact that the temperature scale is universal and independent of the composition of the system.

The first law of thermodynamics deals with conservation of energy. It can be concisely stated in the following form. In an isolated, adiabatic system the total energy U stays constant. In a system that can do work and exchanges heat with its surroundings an energy change dU is the difference between the heat taken up, dq, and the work performed, dW:

$$dU = dq - dW.$$

As mentioned in Chapter 1, Joule, Helmholtz, and others studied the equivalence of different forms of energy and work in the middle of the last century. It has been stressed above (p. 16) that it is the sum of all forms of energy that is conserved. Calorimetric measurements of the energy changes during a chemical reaction can be performed at constant volume in a bomb calorimeter, or at constant pressure in an open vessel. The difference is usually trivial for the kind of system encountered in biochemical investigations, but some comments are required to define certain terms.

When the volume change $\Delta V = 0$, then no work is performed and $p\Delta V = 0$. In such cases

$$\mathrm{d}U = \mathrm{d}q. \tag{2.4}$$

If, however, the reaction is also used to perform work by moving a piston, then

$$\mathrm{d}U = \mathrm{d}q - p\mathrm{d}V. \tag{2.5}$$

The total heat content of a system is called its enthalpy H. The total increment in heat content when the volume changes at constant pressure is

$$\mathrm{d}q = \mathrm{d}U + p\mathrm{d}V = \mathrm{d}H. \tag{2.6}$$

In integral form enthalpy can be defined by

$$H = U + pV$$

and in the differential form by

$$\mathrm{d}H = \mathrm{d}U + p\mathrm{d}V + V\mathrm{d}p. \tag{2.7}$$

At constant pressure equation (2.7) reduces to equation (2.6), which is a condition normally encountered in biochemical reactions.

Internal energy U and enthalpy H are state functions, which means that the change in H, for instance when a system passes from state A to state B, is independent of the path taken in the transition:

$$H_{\mathrm{B}} - H_{\mathrm{A}} = \Delta H.$$

The symbol q is reserved for the heat measured during a calorimetric experiment. The enthalpy change during transformation of a system from state A to state B, at constant T and P, is defined by ΔH:

$$q = H_{\mathrm{B}} - H_{\mathrm{A}} = \Delta H.$$

It should be noted that q and ΔH are positive when heat is absorbed by the system during the process (endothermic).

If one is concerned with processes involving gases, a major concern in many areas of physical chemistry, a much more detailed analysis of the effects of volume changes on the energy balance is required (see for instance Kauzmann, 1967, Chapter 2). Some problems connected with volume changes and the effects of pressure on equilibrium processes in solutions are discussed in Section 4.11.

The second law of thermodynamics is often considered to be conceptually more difficult than the first because it has far-reaching philosophical implications and it requires more precise statements about the boundaries of the system than are sometimes given.

The second law is concerned with restrictions in the direction of energy flow in nature which are not envisaged in the first law. The second law was formulated to take account of the irreversibility of natural processes, a ubiquitous phenomenon. Heat flows from hotter to colder bodies, but not in the reverse direction. Electrical or mechanical energy can be completely converted into heat through electrical resistance or mechanical friction, but heat engines can convert thermal energy only partially into work. If two different gases are allowed to mix, the mixing process will proceed spontaneously, but the mixture will not separate spontaneously into the two pure components (however, see, p. 31). To achieve this, if indeed it can be done at all, work must be done. Every natural process is to some extent irreversible, although we can observe processes in which an infinitesimal change of conditions can cause a process to shift reversibly in one direction or the other. For instance, in an ice–water mixture at 273.15 K the two phases are in equilibrium and reversible melting and freezing occurs. A very slight rise or fall in temperature can also still maintain reversibility. Similarly, if a battery is discharged at a finite rate, some irreversible changes occur and the battery cannot deliver its maximum voltage. However, if the voltage of the battery is counterbalanced by an equal external voltage, as in a potentiometer, the flow of the current is zero and the voltage of the battery corresponds essentially to the maximum attainable under reversible conditions. Such reversible processes provide essential information for the application of the second law of thermodynamics.

The essential quantity in nature, in terms of which the second law is defined, is entropy, denoted by the symbol S. Unlike energy, which is conserved according to the first law, entropy is not conserved, but increases with time in any spontaneous process in an isolated system.

In the limiting case of a reversible process, the change in entropy of a system is given by the heat absorbed by the system divided by the absolute temperature. Thus when ice melts it absorbs 1440 cal, or 6025 J, per mole of H_2O and the entropy change, if the process occurs at constant pressure, is

$$\Delta S = \frac{\Delta H}{T_m} = \frac{1440}{273.15} = 5.27 \text{ cal K}^{-1}\text{ mol}^{-1} = 22.06 \text{ J K}^{-1}\text{ mol}^{-1},$$

where ΔH is the heat absorbed and T_m the melting temperature. More examples and a detailed rationale for the above expression for entropy will be found later in this chapter.

The thermodynamic temperature T, which is involved in the definition of entropy, is in fact identical with the temperature determined by a perfect gas thermometer, by the procedure we have outlined. Proofs of this proposition are given in most treatises on thermodynamics – see for instance Edsall and Wyman (1958, pp. 146–159) or Lewis and Randall (1961, pp. 83–85). Here we shall not present the proof, but simply accept the correspondence of the two temperature scales as an established fact.

Another type of process that can, in principle, be carried out reversibly, consists of heating (or cooling) a substance over a temperature range, say from T_1 to T_2. This can be done by transfer from a heat reservoir to the substance, the temperature of the reservoir being slightly above that of the substance at each stage. A slight decrease in the temperature of the reservoir would result in heat flow in the other direction. If the process proceeds at constant pressure, the heat absorbed (dH) during an infinitesimal temperature rise (dT) is the heat capacity $(\partial H/\partial T)_p = C_p$ (the subscript indicates constant pressure). It follows that

$$dH = C_p \, dT \tag{2.8}$$

and

$$dS = C_p \frac{dT}{T} = C_p d(\ln T). \tag{2.9}$$

We obtain the total entropy change between T_1 and T_2 by integration:

$$\int_{T_1}^{T_2} C_p \, d(\ln T) = S(T_2) - S(T_1). \tag{2.10}$$

In general C_p is a function of temperature, but over a limited temperature range it is justified to assume, in approximate calculations, that it is constant:

$$\Delta S = C_p \int_{T_1}^{T_2} d(\ln T) = C_p \ln (T_2/T_1). \tag{2.11}$$

We may take as an example of an irreversible process the mixing of 1 mole of cold water at 273.15 K with 1 mole of water at 373.15 K. If we assume for simplicity that C_p is constant over this temperature range, we shall finish with 2 mole of water at 323.15 K. This irreversible mixing can be replaced by two reversible processes, which together lead to the same result. We can warm the cold water reversibly and cool the hot water reversibly. The sum of the two processes gives an increase in entropy:

$$\Delta S = C_p \ln \frac{323.15}{273.15} + C_p \ln \frac{323.15}{373.15} = C_p(0.1681 - 0.1439)$$
$$= C_p \times 0.0242.$$

This result can be generalized for mixing two liquids of temperature T_1 and T_2:

$$\Delta S = C_p \ln \frac{(T_1 + T_2)^2}{4T_1T_2} \cdot \tag{2.12}$$

We can readily show that the logarithmic term is positive:

$$(T_1 + T_2)^2 = (T_1^2 + 2T_1T_2 + T_2^2) > 4T_1T_2$$

and when $4T_1T_2$ is subtracted from both sides of this inequality it becomes

$$(T_1^2 - 2T_1T_2 + T_2^2) = (T_1 - T_2)^2 > 0.$$

This is obviously true since a squared term must be positive.

The relation $dS = dH/T$ holds only for reversible processes. However, we have shown the equivalence of entropy changes via reversible and irreversible pathways. S, like U and H, is a property of a system – a state function. When the conditions are specified, the value of ΔS, for the transition of the system from state 1 to state 2, depends only on the initial and final states and not on the path taken to go from one to the other. If we can find a reversible way to get from state 1 to state 2, we know that the entropy difference by another irreversible path will be the same. This has been illustrated above. However, if the path taken is irreversible, the surroundings of the system must undergo an increase in entropy that would not occur if the process were reversible.

Two other examples of entropy calculations may be found instructive. The boiling point of water at 1 atm (101 325 Pa) is 373.15 K and the heat of evaporation per mole at that temperature and pressure is 40.88 kJ. The entropy of evaporation is

$$\Delta S = \Delta H/T = +109.55 \text{ J K}^{-1} \text{ mol}^{-1}.$$

A similar calculation for *n*-hexane (whose boiling point at 1 atm is 341.85 K, and whose heat of evaporation is 28.853 kJ mol^{-1}) gives

$$\Delta S = \Delta H/T = +84.40 \text{ J K}^{-1} \text{ mol}^{-1}.$$

The reason for the greater increase in entropy for the phase change from liquid to vapour for water as compared with that for hexane is the fact that liquid water has a more ordered structure due to hydrogen bonding. We shall return to this explanation when the connection between order and entropy is discussed.

Another thermodynamic function of great importance is the Gibbs energy G, named after Willard Gibbs, who first used it in his great treatise published in 1876. The key role of this function will be apparent in the discussion of several topics in this volume. Here we introduce it to show its relevance to the second law.

The Gibbs energy is defined by the relation

$$G = H - TS = U + pV - TS.$$

This function is particularly useful in analysing processes taking place at constant pressure and temperature. These conditions are commonly a close approximation of biological processes. If we differentiate G, we obtain

$$dG = dH - T\,dS - S\,dT.$$

At constant T the last term drops out, and for macroscopic changes in enthalpy and entropy we can also write

$$\Delta G = \Delta H - T\Delta S. \tag{2.13}$$

The condition for a spontaneous process to occur, at constant p and T, is that ΔG must be negative. Such a process is termed exergonic, by analogy with the term exothermic (see p. 21) which denotes a negative ΔH. Endergonic processes, in which ΔG increases, do not occur in nature unless they are coupled with exergonic processes that supply a sufficient decrease in Gibbs energy to make ΔG for the entire combined process negative. Such coupling occurs constantly in biochemical systems. The hydrolysis of ATP, for instance, is exergonic and through suitable coupling can drive many reactions which, by themselves, could not occur. Such processes will be described in some detail in Chapter 4, and other linked processes occurring in ligand binding will be illustrated in Chapter 5. It will be seen later that a considerable part of the information available on thermodynamic quantities is derived from changes in the Gibbs energy during chemical reactions: ligand binding, macromolecular assembly, and other processes which can be studied under equilibrium conditions.

The third law of thermodynamics evolved from the 'heat theorem' that was first formulated by Walther Nernst in 1906. This concerned the fact, established for many systems, that the values of ΔH and ΔG approach one another more and more closely at very low temperatures, and the difference $\Delta H - \Delta G$ appears to extrapolate to zero as $T \to 0$ K. This also requires that the change in heat capacity (ΔC_p) in such processes must fall rapidly to zero under the same conditions. Since $\Delta H - \Delta G$ approaches zero, this requires from equation (2.13) that ΔS also goes to zero for such processes at 0 K. Nernst recognized that, in view of these facts, it should be possible to calculate the values of chemical equilibrium constants from thermal data – heat capacities, heats of transition (such as melting), and heats of reaction. This was in general impossible on the basis of the first two laws alone.

In 1911 Planck went beyond Nernst's formulation, and stated the third law in essentially the form in which it is still given today: namely that the entropy of any perfect crystal could be taken as zero at 0 K. This meant that absolute values of entropy for all such substances could be determined from thermal measurements, as we shall explain below.

To make practical use of Planck's statement of the third law, we must have experimental criteria to judge whether a given crystal can be considered perfect at very low temperatures. A perfect crystal should have a completely ordered structure at 0 K; an imperfect crystal must be partially disordered. To discriminate between these possibilities we must go beyond classical thermodynamics and obtain evidence concerning crystals at the molecular level. Such a familiar crystal as ordinary hexagonal ice actually retains some residual disorder, even at the lowest temperatures. We discuss later in this chapter how the entropy due to this disorder can be calculated. Glasses, and non-crystalline solids in general, contain residual disorder and the third law does not apply to them.

For a substance that can be considered a perfect crystal at temperatures approaching 0 K, the third law permits determination of its absolute entropy.†

†In denoting entropy values as 'absolute', we almost always accept the convention that a mixture of isotopes is to be regarded as a single substance. Thus sodium chloride crystals can be considered as perfect, and therefore their entropy is taken as zero, at 0 K. However, the chloride

In fact, absolute zero is unattainable – this is, indeed, one way of stating the third law, but experimentally it is possible to approach it so closely that a reliable extrapolation to 0 K is possible.

Assuming the validity of the third law, we can now integrate the function C_p/T from $T = 0$ up to the first point at which a phase transition (usually melting) takes place at a temperature T_m:

$$S = \int_0^{T_m} \frac{C_p}{T}\, dT. \tag{2.14}$$

The phase transition takes place at a constant temperature, with absorption of heat ΔH_m, and the entropy increase is

$$\Delta S_m = \Delta H_m / T_m. \tag{2.15}$$

If we continue to heat the substance above the melting point to a higher temperature T_2, the further entropy increase is

$$\Delta S(T_m \rightarrow T_2) = \int_{T_m}^{T_2} \frac{C_p}{T}\, dT = \int_{T_m}^{T_2} C_p\, d(\ln T). \tag{2.16}$$

If further phase transitions occur (such as vaporization), the entropy change is treated just like that for T_m, and the total entropy of the substance is given by the sum of the various terms in the equations above.

The application of specific heat measurements to the study of phase transition and conformation changes in macromolecular systems will be discussed in detail in Section 6.2.

It should be noted that, for the third law to be valid, C_p must fall very rapidly as T approaches 0 K since the entropy integral (equation (2.16)) must be finite, but ln T goes to $-\infty$ as $T \rightarrow 0$. If the integral is to be finite C_p/T must go rapidly towards zero as $T \rightarrow 0$; this is in fact the case.

2.4 Entropy: phenomenological and statistical basis

It was stated earlier that the laws of thermodynamics are laws of experience. Experience tells us that if we have two vessels with water at the same temperature in thermal contact, we are not likely to find a significant temperature difference between them arising spontaneously. Such an event – water freezing in one vessel and boiling in the other – need not contradict the first law of thermodynamics. The total amount of energy could be constant. However, the second law of thermodynamics would be violated because the entropy of the whole system, consisting of the two vessels, would decrease spontaneously:

ions are a mixture of $^{35}Cl^-$ and $^{37}Cl^-$ in a ratio of about 3 : 1. Thus, if we discriminate between isotopes, a crystal of pure $Na^+\ ^{35}Cl^-$ would be more perfect, and have lower entropy at 0 K, than a crystal of ordinary $Na^+\ Cl^-$. By convention, however, we neglect this fact in all researches in which isotope separation is not involved.

$$\Delta S = \frac{\Delta H}{T_1} - \frac{\Delta H}{T_2} < 0. \tag{2.17}$$

T_1 is the higher temperature at which heat is taken up and T_2 the lower temperature at which the same amount of heat is released.

Similarly, experience tells one the outcome of the following 'thought experiment'. Imagine a vessel with a barrier in the middle and gas on one side and a vacuum on the other side of the barrier. When the barrier is lifted the gas will gradually distribute itself homogeneously among all the volume elements of the system. On the other hand, if the vessel is filled with gas and the barrier inserted at intervals it is very unlikely that one will find all the gas molecules on one side of the barrier. Just how unlikely will be discussed below.

The example of heat transfer relates to the classical description of entropy, while the distribution of gas among volume elements relates to the statistical description. The connection between heat flow and the tendency of a system to proceed towards a random distribution of the maximum number of states (spontaneous processes) was used by Boltzmann to formulate the statistical laws of entropy. The statistical interpretation has not only helped to illustrate thermodynamic behaviour with molecular models, it has also led to the use of entropy as a characteristic of other than molecular systems where statistical considerations govern the behaviour. A recent book entitled *The Entropy Law and the Economic Process* (Georgescu-Roegan, 1971) illustrates many aspects of the application of entropy as a characteristic of social systems.

Shannon and Weaver (1949) and also Szilard (1925) have proposed a relation between entropy and information. To be informative, a message must have a pattern of order. If the message is garbled so that the meaning is partly or wholly lost, information has disappeared and order has been replaced by disorder. There is an obvious relation between this process and what happens (for instance) when a solid melts and the orderly crystal structure is replaced by the more random motion of the molecules of the liquid. This has led the above authors to propose that the production of information is in effect the production of negative entropy (increase in order). In spite of the obvious analogy between the two kinds of processes, the attempt to embody this general idea in quantitative terms runs into serious difficulties and Popper (1976) and others hold it to be invalid.

One important reason why entropy is commonly regarded as a concept which is difficult to understand is the fact that so many ill defined, general, and philosophical statements are made about it. The use of this concept in such diverse disciplines as cosmology, evolution, information, economics, and physical, chemical, and biological laboratory experiments requires very rigorous definitions about the systems, their boundaries, environments, and parameters. The statement that 'entropy always increases in a spontaneous process' requires qualification on two accounts. First one has to define the type of system within which the process occurs – it holds in general only for an isolated system – and secondly one has to consider the stochastic nature of entropy. This latter point will be taken up in Section 2.5. At 90 °C and a pressure of 1 atm steam will condense to water spontaneously and the change from steam to water results in a decrease in entropy, when considered in isolation. However, if we consider this process to occur in an adiabatic system (see p. 10) we find that the heat of condensation per mole, −40.88 kJ, will (a) raise the temperature of the system and (b) bring the system to equilibrium. If the process occurs in a closed isothermal system, the heat removed will increase the entropy of the environment. As a result the total entropy change will be given by

$$\Delta S_{\text{total}} = \Delta S_{\text{system}} + \Delta S_{\text{surroundings}} \geqslant 0.$$

Other examples of ill defined statements are 'the energy of the world is constant', 'the entropy of the world tends towards a maximum', and 'the formation of complex structures from protein molecules will result in an ordered structure and a decrease in entropy'. One must ask oneself about the 'boundaries of the world' which result in an adiabatic, closed system and about the reactions associated with the formation of a complex structure. In fact the assembly of some virus molecules from their subunits, which in itself clearly involves an increase of order, nevertheless results in an increase of entropy in the system as a whole. This is due to the liberation of water of solvation from the components and the consequent gain in rotational and translational entropy of the solvent (see Section 2.6).

2.5 Entropy in terms of molecular statistics

Before discussing examples of the contribution of entropy to some selected chemical and biological events we shall define it in terms of molecular statistics. As pointed out above, Boltzmann suggested a relation between two types of spontaneous unidirectional processes; the flow of heat from a hotter to a colder body and the tendency of a system towards greater disorder. The order of a system is associated with the number of possible states. Obviously if a system exists in a single state the probability of finding it in that state is 1. Although Boltzmann and Planck (for an historical account see Flint, 1958) said that entropy is a function of W (the German word for probability is *Wahrscheinlichkeit*), they used this symbol to define the number of microscopic configurations or states of the system that are compatible with its macroscopic state. This is the reciprocal of the probability of finding the system in one of the states. Therefore W goes from 1 to large numbers, while probability goes from 1 to 0. Now the question arises, what kind of function is

$$S = f(W)?$$

It is shown on p. 40 that entropy is an extensive property, therefore the combined entropy of two systems A and B is

$$S_{\text{AB}} = S_{\text{A}} + S_{\text{B}} = f(W_{\text{A}}) + f(W_{\text{B}}). \tag{2.18}$$

However, combined probabilities result in a product and the number of configurations is

$$W_{AB} = W_A W_B.$$

From this reasoning Boltzmann and Planck formulated the logarithmic relation

$$f(W) = k \ln W,$$

where k is a constant which will be discussed below. It is named the Boltzmann constant.

Another link between the classical and the statistical interpretations can be considered as follows. When heat is taken up (reversibly) at constant temperature,

$$q/T = \Delta S,$$

more states of the system are populated, that is more configurations are possible.

Obviously, even the simplest chemical system will have an enormous number of configurations (electronic, rotational, and translational), making calculations of total entropy a complex matter. However, two simplifications apply to problems which one would normally wish to attempt in connection with biological processes. First, we are not going to concern ourselves with quantum statistics (Kauzmann, 1967; Moore, 1963; Atkins, 1978), which only become significant at the high frequencies of electronic and vibrational energy levels. Secondly, we are concerned not with total entropy but with specific entropy changes during a defined process such as a phase transition or the stretching of an elastic fibre. Examples of such phenomena are descibed in Sections 3.5, 3.28, and 6.3. Let us, however, go back for a moment to the classical and statistical treatment of the expansion of an ideal gas, which incidentally can equally well be applied to diluting or mixing solutions. This will also help to introduce combinatorial algebra, which is required to evaluate the number of possible configurations. In an isothermal reversible (very small) volume change dV we can obtain the entropy change of expanding a gas from volume V_1 to V_2:

$$\mathrm{d}H = p\mathrm{d}V = nRT\frac{\mathrm{d}V}{V}; \tag{2.19}$$

thus

$$\frac{\mathrm{d}H}{T} = nR\frac{\mathrm{d}V}{V}; \quad \Delta S = \int_{V_1}^{V_2} \frac{\mathrm{d}H}{T} = \int_{V_1}^{V_2} nR\frac{\mathrm{d}V}{V} = nR \ln \frac{V_2}{V_1}. \tag{2.20}$$

For the statistical treatment of the same problem we proceed as follows. If a gas expands isothermally the change in the number of configurations depends only on the increase in volume elements available to the same number of gas molecules. The number of spatial configurations available to a single molecule is proportional to the volume V and of N molecules to V^N. Therefore

$$\Delta S = k \ln \left(\frac{V_2}{V_1}\right)^N = nR \ln \frac{V_2}{V_1}, \tag{2.21}$$

where $R = Nk$.

Let us now introduce combinatorial algebra to demonstrate that a uniform distribution of gas molecules results in the largest entropy; the same reasoning applies to the distribution of solute molecules in a solution. We are concerned with fluctuations of local concentrations and not with configurations within the available space as above. If we divide the system into n volume elements and the total number of molecules N_0 is distributed among these elements with N_i molecules in the ith element,

$$\sum_{i='}^{n} N_i = N_0,$$

then the number of possible combinations is given by

$$W = \frac{N_0!}{N_1!N_2!N_3! \ldots N_n!} \cdot \tag{2.22}$$

(see for instance Rushbrooke, 1949, p. 322). Taking the example of 10 molecules distributed among five volume elements, it is readily seen that W is at its maximum when each element contains two molecules. This can be generalized to show that W is always at its maximum when all N_i are equal:

$$N_i = N_0/n.$$

This leads to the common sense conclusion that the homogeneous distribution is the most probable one. The most probable state is the one which can be achieved in the largest number of different ways. There are some logical difficulties in defining and distinguishing states of identical molecules, which have been discussed recently by Braunstein (1969). The 'Gibbs paradox' illustrates the problem. The use of equation (2.21) for the number of configurations of a pure gas and consequently the dependence of entropy on volume (equation (2.22)) leads to

$$S = nS_0 + nR \ln V,$$

where S_0 is the entropy of the standard reference state. If the volume is halved by a partition, the combined entropy of the two halves of the system is

$$S' = 2\left[\frac{n}{2} S_0 + \frac{n}{2} R \ln \frac{V}{2}\right] = n\, S_0 + n\, R \ln \frac{V}{2} \cdot \tag{2.23}$$

This, clearly, does not make sense. The paradox arises from the fact that for derivation of equation (2.23) the molecules were assumed to be distinguishable. Since the molecules in a pure gas are not distinguishable, the entropy change of putting up or removing a barrier is zero. This problem is discussed by Kauzmann (1967, p. 231) and Rushbrooke (1949, p. 177) and their reasoning for correcting this paradox should be consulted. It leads to the Sackur–Tetrode equation which relates the entropy of a monatomic gas to the specific molecular properties of the gas.

At equilibrium the entropy is at its maximum and one expects the most probable distribution. However, equilibrium is a dynamic situation involving continuous fluctuations. The existence of local fluctuation in the pressure or density of a gas revealed by the Rayleigh scattering giving the blue colour to the sky, the Shot effect in electronic devices, and the Brownian motion observed when colloidal particles are viewed under a microscope are due to this phenomenon. It will be seen that the relative magnitudes of the fluctuations depend on the number of molecules involved. A small particle is in collision with relatively few solvent molecules and the collisions in different directions are therefore not uniform at any one instant in time.

It will be seen that fluctuations can become important in biological systems in a variety of ways. It is therefore worthwhile to make a brief diversion to discuss this phenomenon.

The common sense experience, on which the second law of thermodynamics is based, tells us that in a vessel filled with a gas at equilibrium we will *never* find all the gas molecules in one half of the volume. The statistical interpretation of entropy is less definite because it assigns a probability to such an event.† Each molecule of gas has the probability 0.5 of being in a particular half of the volume. For 1 mol of gas the probability of all 6.03×10^{23} molecules being in this particular half of the volume is

$$0.5^{6.03\times10^{23}} = 1/10^{1.8\times10^{23}}.$$

This means that if one checked the distribution once every second for $10^{1.8\times10^{23}}$ s the probability of finding such a distribution once is 0.634. It is worth noting that the age of the Earth is $\simeq 1.5 \times 10^{17}$ s. While this is not a very likely event, can one really say never? This is where thermodynamics and philosophy come together. However, it will be seen that less extreme fluctuations can be evaluated and are of experimental significance.

It is of interest to consider what happens if one makes observations of the state of the system at a greater frequency than $1\ s^{-1}$. In this way one could

†If the probability against an event occuring is 100 : 1, then the probability of the event occurring once in 100 times is 0.63. The general expression for the case of a probability against an event occurring being n : 1 is given by P_n, the probability that it will occur once in n trials:

$$P_n = 1 - \left(\frac{n-1}{n}\right)^n$$

the limiting value for P_n as $n \rightarrow \infty$ is $1 - e^{-1}$.

find a particular distribution in a shorter time. However, quite apart from experimental limitations of the frequency of observation, there is also a theoretical one. The kinetic behaviour, i.e. the velocity of the gas molecules, determines for how long there is a correlation between an observed distribution and subsequent ones. If the interval between two observations is too short, the observed distributions would be correlated; that is they would no longer be independent of each other. When significant fluctuations occur, the decay to the mean value has a relaxation time characteristic of the kinetic processes. Examples of applications of this phenomenon are given below.

A probability distribution designed to estimate the number of events occurring in a segment of a continuum of time, length, or space is given by the Poisson equation. This can be applied to fluctuations of the number of particles in a volume segment of a continuum under random movement (at equilibrium, without interactions). If the average number of particles per segment is z, the probability that x particles will be found in a given segment is given by

$$P_x = \mathrm{e}^{-z}\, z^x/x!$$

Some assumptions and conditions are made in the derivation of this equation (see, for instance, Kauzmann, 1967, p. 208). The great advantage of this distribution function over others (binominal or Gaussian) is that only the average number of particles or events per unit segment of space or time need be known to calculate the probability of a given deviation from this mean.

It can readily be calculated from the above case of two particles per volume element as average population that the probability of doubling the population per element is 0.09 or a 9% chance. If one increases the population to an average of 30 particles, the probability of finding twice the average number per volume element is 4.8×10^{-7}, or less than one in a million.

The above considerations have practical consequences for the analysis of random events such as counting radioactive disintegrations, photons, or moving particles within a grid. The larger the number of counts, the smaller the percentage fluctuation from the mean. The probability of a fluctuation $\pm\sigma$ from the mean is 2/3 when

$$\sigma = \frac{\sqrt{n}}{\text{unit}}$$

where n is the number of counts or events and the units can be a given time interval, volume element, or area. In this way one can reduce random noise from experimental measurements by repetitive procedures.

Another aspect of fluctuations or noise has already been referred to, namely that they are related to the kinetic properties of the system. This interesting topic is beyond the scope of the present volume, but it should be pointed out that its application to the study of the dynamic behaviour of macromolecules

(Rigler and Ehrenberg, 1976; Ehrenberg and Rigler, 1976) and physiological systems (Stevens, 1977) has become a potentially powerful tool. The rate processes in a complex system are reflected in the frequency spectrum of the fluctuations. As pointed out above, if one is able to take ever more rapid pictures of the distribution of gas molecules one can determine the time scale on which there is a correlation and therefore get information about the rate of movement. Laser light scattering observations are making this possible.

Thermodynamic properties depend on the statistical behaviour of a system. A molecule or even a thousand molecules in a vessel do not have a temperature or pressure. These intensive parameters are independent of the size of the system as long as it contains enough molecules for local fluctuations in numbers and average kinetic energy to be negligible. Extensive properties such as energy, enthalpy, and entropy are dependent on the size of the system. If two volumes of the same gas at the same temperature and pressure are combined by removing a barrier between them, the temperature and pressure will remain the same, but the enthalpy, etc., will be additive (see also section 3.2.).

Temperature is a measure of the average translational kinetic energy of the molecules in a system:

$$\tfrac{1}{2}\, mv^2 = \tfrac{3}{2}\, kT.$$

In complex molecules, as the total energy is increased, a smaller increment in mean kinetic translational energy, and hence in temperature, occurs since the added energy is divided among a larger number of degrees of freedom (rotational, vibrational, etc.). Two systems have the same temperature if they remain in equilibrium when placed in thermal contact with each other. Temperature can be measured as a length, as a pressure, or in terms of many other quantities.

The difficulties in defining temperature in a gas at very low pressure do not occur in a solution containing only very few solute molecules since the kinetic energy will be equilibrated by the solvent, which defines the temperature. In biological processes one often finds situations in which only small numbers of a particular molecule may occur within a cell or a subcellular particle. This applies, for instance, to some effectors and messengers in genetic and metabolic control. There are, for instance, only about 10 molecules of *lac* repressor protein in a single cell of *E. coli*. In such cases one has to reach conclusions from integrations over time rather than numbers. If one considers the degree of occupation of regulatory sites by effector molecules one can, in the case of small numbers, define the proportion of the time a site is occupied instead of using the definition of proportion of total sites occupied.

The so called 'ergodic hypothesis' attempts to establish the equality of time averages and ensemble averages. The word hypothesis rather than theorem is used deliberately because the general proof of this problem in statistical mechanics is under discussion at a high level (see for instance Margenau and Murphy, 1943, p. 427; Georgescu-Roegen, 1971, p. 153).

2.6 Residual entropy and properties of water

Some simple examples of entropy calculations and comparisons with experimental measurements may be helpful in illustrating certain phenomena. Interesting features have been discovered when experiments and calculations disagreed because the assumed model was too simple.

One simple example of such discrepancies is residual entropy at 0 K. For instance, it was found that specific heat measurements on crystals of carbon monoxide extrapolated to 0 K indicated a residual entropy of 4.2 J K^{-1} mol^{-1}. If one postulates that positions in the crystal lattice can be taken up by molecules in two orientations, C—O and O—C, then there are two choices for each location. N locations per mole of CO give 2^N choices for the distribution of C—O and O—C. From Boltzmann's formula,

$$S = k \ln 2^N = R \ln 2 = 5.76 \text{ J K}^{-1} \text{ mol}^{-1}.$$

From the higher value for the calculated, as compared with the experimental, value, one can conclude that the distribution between the two orientations is not quite random.

A similar result is obtained in other situations in which some disorder is 'frozen in'. This can be due to the fact that solidification is not a slow enough process.

Water has a number of distinctive properties due to its dipolar structure. Because of the great importance of water in influencing the behaviour of molecules dissolved in it, some of its thermodynamic properties deserve detailed attention. It has already been mentioned (p. 24) that liquid water has a considerably more ordered structure than other liquids. The transition of ice to water and to vapour gives useful information about the amount of organization in the liquid state. The heat of sublimation of ice (51.1 kJ mol^{-1}) gives the heat required to separate water molecules from their positions in ice crystals to that in water vapour. The van der Waals attraction between water molecules is estimated at about 11.6 kJ, which leaves about 38 kJ for breaking two hydrogen bonds per water molecule on vaporization from the crystalline state. The fusion of ice, that is the transition from the crystalline to the liquid state, requires only 6 kJ mol^{-1}. This indicates that only 16% of hydrogen bonds are broken during melting and that water molecules in the liquid have a considerable amount of local order.

While there is considerable local order in liquid water, there is a movement of protons in the ice crystal lattice, resulting in residual entropy at 0 K. The distribution of hydrogen bonds and covalent bonds in an ice crystal can be represented in the following way:

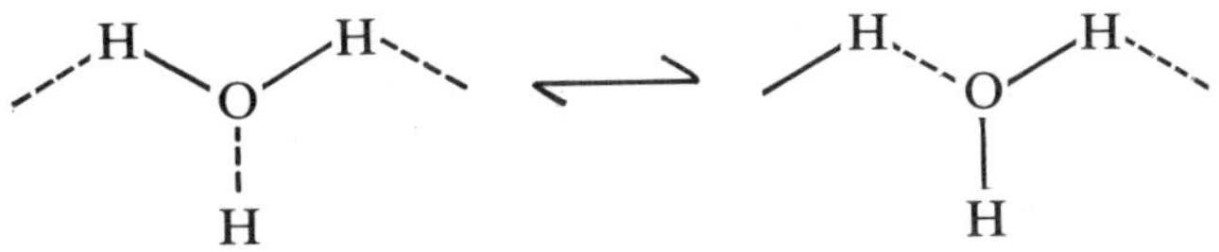

The hydrogen bond length O----H is 0.177 nm and the covalent bond length is 0.099 nm. The hydrogen nuclei can jump from one type of bonding to the other as indicated above. The frequency of this, the fastest known unimolecular reaction, is 10^{11} s^{-1}. In 1 mole of ice there are N oxygen atoms and $2N$ hydrogen nuclei. If each hydrogen nucleus can be in either bonding position, $2N$ nuclei with binary choice would have 2^{2N} possible configurations. However, there is the restriction that each of the N oxygen atoms can only be covalently bonded to two out of four surrounding H nuclei. The number of possible configurations is restricted by $(6/16)^N$, because without the restriction each of the N oxygens could bond in 16 ways (two alternatives in each of four directions) while with the restriction of covalent combination in four directions, two at one time,

$$ {}_4C_2 = \frac{4!}{(4-2)!2!} = 6. $$

The total number of configurations due to the mobility of protons is $(6/16)^N \times 2^{2N}$ and

$$ S = R \ln (3/2) = 3.37 \text{ J K}^{-1} \text{ mol}^{-1}. \tag{2.24} $$

Experimental measurements of the residual entropy of ice crystals extrapolated to 0 K give $S = 3.43$ J K^{-1} mol^{-1}, which supports the validity of the proposed model.

The characteristic properties of water and the structural features which cause them have interesting consequences. The dipolar structure of liquid water results in an exceptionally high dielectric constant (78.54 at 25 °C as compared to 2.2 for cyclohexane). The energy of interaction between electric charges is inversely proportional to the dielectric constant of the medium (see Section 5.3). Macromolecules, and especially proteins, have specific polar and non-polar regions and their exposure to solvent or ligand molecules can be modulated by conformation changes. This results in a controlled competition between intramolecular forces and interaction with water as well as in a control of the local dielectric constant on the exposed surface.

An important physiological consequence of the properties of water is its suitability as a thermostat of biological systems. Its high specific heat means that a lot of energy is required to raise the temperature of water (4.184 J to increase the temperature of 1 g of water by 1 °C). This is only surpassed among liquids by ammonia (5.15 J K^{-1}). Also, the large heat of vaporization of water (see p. 24) results in efficient cooling through perspiration and consequent evaporation. The heat lost for each gram of water evaporated is

2.84 kJ. Henderson (1913) and, in a modern version, Edsall and Wyman (1958) give more detail about the role of water in biological phenomena. Henderson coined the apt phrase 'the fitness of the environment' to describe the biological significance of the properties of compounds of carbon, hydrogen, and oxygen, with special emphasis on water.

Apart from the effect of water on the electrostatic forces in solutions mentioned above and the obvious competition between intramolecular and solvent–solute hydrogen bond formation in macromolecules, there is another important contribution of water to the balance of non-covalent forces. The so-called 'hydrophobic effect' is the tendency of molecules containing non-polar groups to cluster together so that the non-polar surface exposed to water is at a minimum. This phenomenon was described by Tanford (1980) as 'perhaps the most important single factor in the organization of the constituent molecules of living matter into complex structural entities such as cell membranes and organelles'. A detailed analysis of the hydrophobic effect would be out of place here. Various models proposed are still being actively disputed. Some thermodynamic consequences should, however, be mentioned. When a given volume of a non-polar liquid is transferred to water, forming a very dilute aqueous solution, the total volume of the system (water plus organic liquid) decreases (Kauzmann, 1959). There is also a negative contribution to the entropy due to changes in the structure of water at the surface of the non-polar (hydrophobic) solute (Tanford, 1980).

Solubility is an equilibrium phenomenon depending on the Gibbs energy change when solute is dissolved. Strictly speaking this equilibrium phenomenon should be expressed in terms of standard Gibbs energy as derived in Section 4.2. With different solutes dissolved in water the two components ΔH and ΔS make different relative contributions to the relation

$$\Delta G = \Delta H - T\Delta S.$$

With ionic solutes relatively large contributions of a decrease in H (negative ΔH) compensate for the decrease in entropy when water becomes more organized around the charged groups. This results in a negative ΔG for such soluble compounds. For hydrophobic compounds ΔH is often small and very temperature dependent and the significant contribution of a decrease in entropy results in a positive ΔG and low solubility. Another important consequence of the entropic nature of hydrophobic effects is the negative temperature dependence of solubility at lower temperatures. ΔG will decrease when a negative entropy term is multiplied by progressively smaller values for T. For a fuller discussion of this complex problem the reader is referred to Section 3.14.

The magnitude of the hydrophobic effect depends on the surface area of non-polar molecules exposed to the aqueous environment. Reynolds, Gilbert, and Tanford (1974) suggest a contribution of 8.5–10.5 kJ mol^{-1} nm^{-2}. Therefore non-polar groups dissolved in water prefer to combine to reduce the exposed surface area – and with that the Gibbs energy of solution. The resulting apparent attractive energy is approximately 3.5 kJ mol^{-1} per —CH_2— unit of a hydrocarbon chain (Smith and Tanford, 1973). For the reasons outlined above, this attraction between hydrophobic areas of molecules increases with temperature. Conversely the assembly of some complex protein structures is sensitive to lowering of temperature and they dissociate in the cold. A more detailed theoretical discussion of this phenomenon will be found in Section 3.14.

2.7 Non-equilibrium thermodynamics

The present volume is concerned only with the design of experiments and the analysis of data on systems in equilibrium. This aspect of thermodynamics provides information about the energetics of isolated chemical processes and changes of state of molecules which form the building blocks of biological systems. The complete systems, however, be they the simplest unicellular prokaryote or even a subcellular organelle, are never in true equilibrium while they are functional. Equilibrium is death to a cell. The branch of thermodynamics dealing with irreversible processes, transport processes, and steady state thermodynamics must be referred to in any treatment of energy relationships in processes of biological significance. The present discussion will be restricted to a statement of some of the basic rules and to references where further valuable information can be obtained. The book by Katchalsky and Curran (1965) and the review by Katchalsky and Spangler (1968) are recommended for further reading on the foundations and applications of this subject.

Equilibrium thermodynamics deals with the behaviour of isolated or closed systems while non-equilibrium thermodynamics is concerned with open systems in which mass transport to or from the environment occurs. A typical and important example is the coupling of diffusion of a reactant into a cell with a chemical transformation and the diffusion out of the cell:

$$A \rightarrow (A \rightleftharpoons B) \rightarrow B.$$

This explains two terms commonly used: transport processes and coupled processes.

The method used to combine these concepts is to define fluxes (J) in terms of a driving force (X) and a constant specific for the particular physical phenomenon. This constant is called the phenomenological coefficient (L). The equation

$$J = XL$$

can be applied to the flow (or flux) of electricity, heat, mass (diffusion), or a chemical process across a gradient of e.m.f., temperature, chemical potential, or chemical affinity. The coefficients $L = J/X$ are expressions of flow per unit force and have the characteristics of conductance or mobility. For any one of these processes to occur in isolation the condition

$$JX \geqslant 0$$

determines the direction. If, however, two or more processes are coupled it is only necessary that the sum

$$\sum_i J_i X_i \geqslant 0$$

points in the right direction.

Examples of coupled processes are, for instance, two chemical processes with affinity (A) as the force, i.e.

$$\Phi = J_{\text{synthesis}} A_{\text{synthesis}} + J_{\text{metabolism}} A_{\text{metabolism}} \geqslant 0;$$

active transport against a concentration gradient and mechanochemical coupling, i.e.

$$\Phi = J_{\text{diffusion}} A_{\text{diffusion}} + J_{\text{metabolisim}} A_{\text{metabolism}} \geqslant 0;$$

and metabolic energy coupled to mechanical work, i.e.

$$\Phi = F\frac{\mathrm{d}l}{\mathrm{d}t} + J_{\text{metabolism}} A_{\text{metabolism}} \geqslant 0.$$

The refrigerator or the Peltier effect are other examples in which mechanical or electrical work is used to pump heat from a colder to a hotter body.

The above statements are insufficient for solving quantitative problems such as coupling diffusion to synthesis or a chemical process. Numerical solutions to questions like ATP hydrolysis coupled to active transport of Na^+ ions against a concentration gradient can be obtained by the methods of Katchalsky and his coworkers (Katchalsky and Curran, 1965; Katchalsky and Spangler, 1968).

Some of the confusions about the application of the first and second laws of thermodynamics to biological systems arise from the literal interpretation of the following statement by Clausius: 'The energy of the world is constant, the entropy of the world tends towards a maximum'. We must refer back to the definition of systems and boundaries in Section 2.1. Without entering into the problems of the thermodynamics of cosmology, let us consider a tadpole in a pond on Planet Earth and within the Solar System:

Solar System → Earth → pond → tadpole.

The entropy of the Earth can decrease and the energy increase, at the expense of an entropy increase and energy decrease on the Sun. The pond can have processes going on which result in $\Delta S < 0$ and $\Delta G > 0$, provided that there are compensating effects on the rest of the planet. Similarly the tadpole can have $\Delta S < 0$ and $\Delta G > 0$ at the expense of the rest of the pond. The increased organization of the growing organism occurs at the expense of disorganization (randomization) somewhere in its environment (see also p. 28 in Section 2.4).

CHAPTER 3

Fundamental concepts: mathematical analysis and introduction to multicomponent systems and chemical potentials

In Chapter 2 we have given a general and impressionistic view of the laws of thermodynamics, and the fundamental thermodynamic quantities that define the systems that will concern us – such quantities as pressure, volume, temperature, energy, enthalpy, entropy, and Gibbs energy. In the present chapter we consider in more detail the quantitative relations among the properties of thermodynamic systems, and develop some of the important mathematical relations between them. In particular we consider in detail the properties of systems composed of variable amounts of different chemical components. The chemical potential of a component, which varies in proportion to its activity or partial pressure, is fundamental in determining the position of physical and chemical equilibrium in any system.

3.1 Components and phases

We begin by considering the meanings of the terms 'phase' and 'component' in thermodynamics. A phase is a region of the system that is homogeneous in a macroscopic sense; that is, if we take a sample of the phase that is large enough to contain a very large number of molecules, the composition of the sample will prove to be the same, no matter what region of the phase we choose in drawing out a sample. A single phase may contain many chemical components, or only one. Pure liquid water is a phase; so is an aqueous solution containing salts, sugars, amino acids, and other substances, at equilibrium. A system of pieces of ice in water is a two-phase system; even if the ice is in many pieces it forms a single phase, since its structure and composition is the same for all of them. Ordinarily there is only one gas phase in any system, since gases diffuse freely into one another. An emulsion of small oil drops in water might be regarded as a two-phase system, but generally it is not a stable system; the oil drops tend to coalesce into a single body of oil above the water, which is indeed a two-phase system. Sometimes it is possible to form an emulsion of oil in water which is stable enough to remain as such for long periods; in such cases the interfacial area between the oil and the water is very large, and the interfacial energy is a significant part of the total energy of the system. We do not consider such cases here; in this chapter, and throughout nearly all of this book, we assume that the phases in the system are defined in terms of pressure, volume, temperature, and composition, and that surface energy is a negligible part of the total energy.

The number of components of a system is the minimum number required to specify its composition precisely. Pure liquid water is a single component, though it contains H^+ and OH^- ions as well as H_2O molecules. The concentrations of ions are fixed under given conditions, so that the component 'water' simply specifies a precise mixture of ions and molecules. A solution of two salts in water – say sodium chloride and magnesium sulphate – is a three-component system. There are four kinds of ions present, but they correspond to only two components, since $[Na^+] = [Cl^-]$ and $[Mg^{2+}] = [SO_4^{2-}]$, where square brackets are used to indicate concentration. A single ion cannot constitute a component, since it cannot be added to or removed from the system independently of other ions. A component, though it may be composed of ions, must be electrically neutral when taken as a whole. Thus a system composed of water and the sodium salt of a pure protein is a three-component system. It may be regarded as being composed of water, isoelectric protein, and enough sodium hydroxide to bring the protein solution to the required pH. Alternatively, in defining the system, we may specify the pH as a third variable, instead of the amount of added NaOH. The pH may thus be regarded as equivalent to one of the components of the system, although this is a less rigorous way of specifying its composition.

3.2 Extensive and intensive properties

Consider a system of n components, at given pressure and temperature. If we are concerned only with its chemical composition and not with its size, we can define the composition in terms of $n-1$ variables by taking one component as the reference component and specifying the amounts of the others relative to it. In aqueous systems water is frequently chosen as the reference component, and the amounts of solutes are given, for instance, as moles of solute per kilogram of water, or in some other convenient unit.

This brings us to the important distinction between extensive and intensive properties. Extensive properties, which include volume, internal energy, enthalpy, entropy, Gibbs energy, and the mass of each of the components, are directly proportional to the size of a system of given composition. Intensive quantities, such as pressure and temperature, are obviously independent of its size. Composition variables, such as moles of component i per mole of component 1, are also clearly intensive. The distinction between these two classes of quantities is important to bear in mind (see also Section 2.5).

3.3 Some mathematical considerations: the role of partial differentiation in thermodynamics

Thermodynamics involves relations between numerous different variables, such as volume, pressure, temperature, and the amount of each of the chemical components in a system. The thermodynamic properties of the system, such as its internal energy (U), its enthalpy (H), its entropy (S), and its Gibbs free energy (G), are functions of these variables. Often we want to know, both in experiment and in calculation, how the properties of the system change when the value of one of the variables changes, while all the others are held constant. Mathematically this corresponds to the process of partial differentiation, which we now consider.

We begin with a simple example. Consider the behaviour of a perfect gas. We can visualize the gas as contained in a cylinder of volume V, which can be changed by moving a piston up or down. The cylinder is surrounded by a bath at an absolute temperature of T kelvins, which can be varied. There is an inlet to the cylinder from a reservoir of the gas, which can be turned off and on, so as to let in more gas, or withdraw some. At the start the cylinder contains n moles of gas. Then the pressure (p) of this ideal gas is given by the familiar equation

$$p = nRT/V. \tag{3.1}$$

We can vary n, T, and V independently. If we vary n, at constant T and V, the rate at which p varies with n is given by

$$(\partial p/\partial n)_{T,V} = RT/V. \tag{3.2}$$

Here the subscripts T and V, attached to the differential expression on the left-hand side, indicate that we are holding these variables constant while varying n. The symbol ∂, preceding p and n on the left-hand side, is used to distinguish a partial differential coefficient from a total differential, which would be written as $\mathrm{d}p/\mathrm{d}n$.

If we vary V, at constant n and T, we obtain

$$(\partial p/\partial V)_{n,T} = -nRT/V^2. \tag{3.3}$$

If we vary T, at constant n and V, we obtain

$$(\partial p/\partial T)_{n,V} = nR/V. \tag{3.4}$$

In each case we treat the quantities indicated by the subscripts as constants during the differentiation. Later, on occasion, it may be useful, for compactness and simplicity of notation, to omit the subscripts, if it is perfectly clear what variables are being held constant in any given equation. If there is any doubt, however, they should always be inserted.

It is important to visualize the meaning of such partial differential coefficients. If for instance we hold n and V constant, as in equation (3.4) above, we can plot p as a function of T. In this case $(\partial p/\partial T)_{n,V}$ is a constant, which defines the slope of a straight line in the p–T plane. Likewise, for fixed values of n and T, the plot of p as a function of V is a rectangular hyperbola (pV = a constant = nRT) and $(\partial p/\partial V)_{n,T}$ is the slope of that hyperbola at any assigned value of V. Thus the partial differential coefficients are *numbers* that give the values of the slopes of such curves at any given point. If we invert one of these coefficients, we simply get a reciprocal number: for instance $(\partial p/\partial V)_{n,T} = 1/(\partial V/\partial p)_{n,T}$. This relation, it should be emphasized, holds only as long as the subscripts – the quantities being held constant – are the same on both sides of such an equation.

Since p is a function of T, V, and n in equation (3.1), we can start from a given state of the system, with specified values of these variables, and vary p by varying each of the other variables in turn. The total infinitesimal increment in p, dp, is then proportional to the increments dT, dV, and dn, of the three independent variables, and the factors of proportionality are the partial differential coefficients:

$$\begin{aligned} \mathrm{d}p &= (\partial p/\partial T)_{n,V}\,\mathrm{d}T + (\partial p/\partial V)_{n,T}\,\mathrm{d}V + (\partial p/\partial n)_{T,V}\,\mathrm{d}n \\ &= R\,[(n/V)\,\mathrm{d}T - (nT/V^2)\mathrm{d}V + (T/V)\mathrm{d}n]. \end{aligned} \tag{3.5}$$

Although equation (3.5) applies strictly only to infinitesimal increments dp, dT, etc., it gives a good approximation if we consider small finite increments. As an example consider 3 mol of a perfect gas at temperature 300 K (26.85 °C) in a volume of 30 l (30 dm^3).

We calculate the pressure of the gas in SI units, with pressure in pascals or kilopascals, and volume in cubic metres. The value of R is thus 8.314 J K^{-1} mol^{-1}, and the volume is 0.03 m^3 = 30 dm^3. Then we find that

$$\begin{aligned} p &= nRT/V = 3 \times 8.314 \times 300/0.030 = 2.494 \times 10^5 \text{ Pa} \\ &= 249.4 \text{ kPa} = 249.4/101.3 = 2.462 \text{ atm} \end{aligned}$$

(1 atm = 101.325 kPa; we use only four significant figures here).†

Suppose that we now increase T from 300 to 301 K, increase V by 3×10^{-5}

†Many text-books of thermodynamics, in making such calculations, express volume in litres (1 litre = 1 dm^3 with high precision) and pressure in standard atmospheres. Then R = 0.082 05 l atm. Thus in the above calculation $p = 3 \times 0.082\,05 \times 300/30 = 2.462$ atm. The use of SI units is just as easy, and has a practical advantage. The work done in compressing a perfect gas from V_1 to V_2 is

$$W = nRT\ln(V_1/V_2)$$

and its numerical value is given directly in joules if we use the SI value of R. No conversion factor is needed, and it is necessary only to use a single value of R (8.314 J K^{-1} mol^{-1}) in all calculations.

One widely used unit of pressure, the bar, equals 100 kPa by definition. The standard atmosphere (760 mmHg) is less than 1.5% greater than 100 kPa; so in calculations of moderate accuracy we may neglect the difference. Another commonly used unit is the torr, essentially identical with a pressure of 1 mmHg. It is important to know the meaning of these units when reading the literature, but we strongly recommend the use of SI units in calculations here and elsewhere.

m^3 = 0.03 dm^3, and increase n by 0.03 mol, the alteration of each of these variables taking place while the others are held constant. Then we have an approximate equation, of the same form as equation (3.5), in terms of small finite increments:

$$\begin{aligned}
\Delta p &= R\,[(n/V)\Delta T - (nT/V^2)\Delta V + (T/V)\Delta n] \\
&= 8.314\left[\frac{3.03}{0.03} - \frac{900}{(0.03)^2}(\times 3\times 10^{-5}) + \frac{300\times 0.03}{0.03}\right] \\
&= 8.314\,[101 - 30 + 300] \\
&= 3084\ \text{Pa} = 3.084\ \text{kPa} = 0.0304\ \text{atm}.
\end{aligned} \tag{3.6}$$

Thus p should increase from 249.4 to 252.5 kPa, or from 2.46 to 2.49 atm. We can of course calculate p directly from equation (3.1), setting T = 301 K, n = 3.03, and V = 0.030 03 m^3. The result, 252.5 kPa, is the same to four significant figures.

We can differentiate partial differential coefficients a second time – or as many times as we please – with respect to the same variable. Thus in the simple case under discussion we find

$$(\partial^2 p/\partial n^2)_{T,V} = (\partial^2 p/\partial T^2)_{n,V} = 0; \qquad (\partial^2 p/\partial V^2)_{n,T} = 2nRT/V^3. \tag{3.7}$$

We may also differentiate first with respect to one of the independent variables, and then with respect to another. Consider for instance the rate of change with temperature, at constant n and V, of the coefficient $(\partial p/\partial V)_{n,T}$, or the rate of change with changing volume, at constant n and T, of the coefficient $(\partial p/\partial T)_{n,V}$. For compactness here we can omit the subscripts indicating which variables are being held constant in each case. We find that these two differential coefficients are equal:

$$\frac{\partial}{\partial T}\left(\frac{\partial p}{\partial V}\right) = \frac{\partial}{\partial V}\left(\frac{\partial p}{\partial T}\right) = \frac{\partial^2 p}{\partial T \partial V} = -nR/V^2. \tag{3.8}$$

In other words, the order of the two successive differentiations does not matter; we get the same result either way, and this is epitomized in the expression $(\partial^2 p/\partial T\,\partial V)$. The order of T and V in the denominator of this coefficient is immaterial.

Equation (3.8) is commonly called the cross-differentiation relation. It is valid for all functions with which we shall have to deal in thermodynamics. In the present case it can readily be verified by the reader for the relations

$$\frac{\partial}{\partial n}\left(\frac{\partial p}{\partial T}\right) = \frac{\partial}{\partial T}\left(\frac{\partial p}{\partial n}\right) = \frac{\partial^2 p}{\partial n \partial T}, \tag{3.8a}$$

$$\frac{\partial}{\partial n}\left(\frac{\partial p}{\partial V}\right) = \frac{\partial}{\partial V}\left(\frac{\partial p}{\partial n}\right) = \frac{\partial^2 p}{\partial V \partial n}. \tag{3.8b}$$

Equations of the type exemplified in equations (3.8), (3.8a), and (3.8b) are of major importance in thermodynamics, and we shall see examples of their use shortly. Some of the important partial differential coefficients are very difficult to obtain experimentally by direct measurement, but we can often show, by equations similar in form to equation (3.8), that they are in fact equivalent to other coefficients that are much easier to obtain by experiment. This is one reason for the immense power of thermodynamics in relating quantities that, at first sight, might seem unconnected.

Similar equivalence relations hold for successive partial differentiations with respect to more than two variables. The final result is the same, regardless of the order in which the successive differentiations are performed.

There is another important relation between partial differential coefficients, which holds when we are considering only two independent variables. We can illustrate this, in the case of the perfect gas, by treating n as constant, so that p is a function only of V and T:

$$\mathrm{d}p = (\partial p/\partial T)_V \,\mathrm{d}T + (\partial p/\partial V)_T \,\mathrm{d}V. \tag{3.9}$$

We may need to know the coefficient $(\partial V/\partial T)_p$, the change of volume with temperature at constant pressure. If we hold the pressure constant, $\mathrm{d}p$ becomes equal to zero in equation (3.9), and in that case we can divide the equation by $\mathrm{d}T$, and rearrange the terms to obtain

$$(\partial V/\partial T)_p = -\frac{(\partial p/\partial T)_V}{(\partial p/\partial V)_T}. \tag{3.10}$$

Here we insert the subscript p on the left-hand side of equation (3.10) to emphasize explicitly that we have imposed the condition of constant pressure for the coefficient $(\partial V/\partial T)$. In this simple case we can obtain the two coefficients on the right-hand side of equation (3.10). The result is very simple:

$$(\partial V/\partial T)_p = V/T.$$

We could have obtained the same result from equation (3.1) directly. However, equations of the type of (3.10), like those of the type of (3.8), often permit us to obtain coefficients that are very difficult to measure experimentally, from a ratio of two other coefficients that may be quite easily measurable.

In this particular illustration of the perfect gas system we chose p as the dependent variable, and T, V, and n as the independent variables. However, in the transformation that led to equation (3.10) we shifted to p as the independent variable, considering it as a function of V and T in the two coefficients on the right-hand side of this equation. We are free in general to choose dependent and independent variables in the most suitable and convenient way for any given problem. We now consider some of these possible choices in relation to some of the major thermodynamic functions: energy (U), enthalpy (H), entropy (S), Helmholtz energy (A), and Gibbs energy (G). First we consider systems of constant composition; later we consider systems in which the amounts of the components present are subject to variation.

3.4 Some partial differential relations for the major thermodynamic functions

The thermodynamic state of a given system in equilibrium at constant chemical composition can generally be specified by fixing the values of two variables. These may be pressure and temperature, or entropy and volume, or volume and temperature, or any of several other possible choices. We assume generally that other classes of variables – for instance the surface energy associated with interfaces in the system – are of negligible importance by comparison. In the study of systems where interfaces are important, of course, we cannot make such an assumption; but for a great many important systems in chemistry and biochemistry we find that the assumption that the system is defined in terms of two independent variables is justified.

From the first law of thermodynamics (see Chapter 2) we know that a small increment ($\mathrm{d}U$) in the internal energy of such a system is equal to the increment in heat absorbed ($\mathrm{d}q$) minus the pressure–volume work done by the system on its surroundings:

$$\mathrm{d}U = \mathrm{d}q - p\,\mathrm{d}V. \tag{3.11}$$

As already pointed out in Chapter 2, the values of $\mathrm{d}q$ and of $p\mathrm{d}V$ in the transition of the system between an initial state (state 1) and a final state (state 2) depend on the path taken in going from one state to the other. Their sum, however, by the first law, is independent of the path and depends only on the initial and final states:

$$U_2 - U_1 = \Delta U = \int_1^2 \mathrm{d}q - \int_1^2 p\,\mathrm{d}V. \tag{3.12}$$

In a system undergoing reversible changes at equilibrium, at temperature T, the change in entropy, $\mathrm{d}S$, equals $\mathrm{d}q/T$; this, like the energy U, is a state function of the system. The change in entropy depends only on the initial and final states. For small reversible variations in a system in equilibrium equation (3.11) therefore becomes

$$dU = T\,dS - p\,dV. \tag{3.13}$$

Thus U is a function of the entropy (S) and the volume (V) of the system. This means that for the increment in energy dU we can also write

$$dU = (\partial U/\partial S)_V\,dS + (\partial U/\partial V)_S\,dV. \tag{3.14}$$

On equating the coefficients of dS and dV in equations (3.13) and (3.14), we obtain at once

$$(\partial U/\partial S)_V = T \quad \text{and} \quad (\partial U/\partial V)_S = -p. \tag{3.15}$$

In fact it is not generally practicable to choose entropy and volume as the independent variables in experimental work, so that these equations are seldom used in practice. It would be extremely difficult, and often impossible, to keep the entropy and volume constant while varying pressure, temperature, or composition of the system.

Pressure and temperature constitute the most practical choice of independent variables for most work in chemistry – and especially in biochemistry. Living organisms are normally under conditions of essentially constant pressure, and mammals normally maintain a constant temperature as well. Almost all organisms can be kept at essentially constant temperature during experimental studies, if this is desired. As we shall see, the choice of temperature and pressure as independent variables leads naturally to the choice of the Gibbs energy, G, as the most useful dependent variable in the study of systems at equilibrium. First, however, we must consider two other important thermodynamic functions.

The first of these is the enthalpy, H. The change in enthalpy, ΔH, in any process taking place at constant pressure is equal to the heat absorbed in the process. This can be seen from the definition of H:

$$H = U + pV.$$

Hence

$$dH = dU + p\,dV + V\,dp. \tag{3.16}$$

If we insert the value of dU from equation (3.13) into (3.16), we obtain

$$dH = T\,dS + V\,dp. \tag{3.17}$$

If p is held constant, we then have $dH = T\,dS$, which is the heat absorbed in a reversible process. In any case, however, H, like U, is a state function of the system, and therefore ΔH, for any process leading from state 1 to state 2, is independent of the path between the two states. Be sure to note that a positive value of ΔH for any process denotes that heat is *absorbed* by the system during that process.

Measurements of processes taking place in constant pressure calorimeters give values of ΔH. There is now a vast array of such calorimetric data that are essential for understanding and predicting biochemical processes (see Chapter 6). Therefore the enthalpy is a function of major importance in biochemical thermodynamics.

The most important partial derivative of the enthalpy is the heat capacity at constant pressure, denoted as C_p.

$$C_p = (\partial H/\partial T)_p = T(\partial S/\partial T)_p. \tag{3.18}$$

It is the change of heat capacity, ΔC_p, in any process, that determines how the enthalpy change, ΔH, varies with temperature at constant pressure. The heat capacity at constant volume, C_V, which is equal to $(\partial U/\partial T)_V$, is also important in certain connections, particularly in the study of gases; but it is C_p that will concern us in this book.

The Helmholtz energy (or Helmholtz free energy), denoted by A, is defined by the equation

$$A = U - TS.$$

Hence

$$\mathrm{d}A = \mathrm{d}U - T\,\mathrm{d}S - S\,\mathrm{d}T. \tag{3.19}$$

Inserting the value of $\mathrm{d}U$ from equation (3.13) we obtain

$$\begin{aligned}\mathrm{d}A &= -p\,\mathrm{d}V - S\,\mathrm{d}T \\ &= (\partial A/\partial V)_T\,\mathrm{d}V + (\partial A/\partial T)_V\,\mathrm{d}T.\end{aligned} \tag{3.20}$$

Hence

$$-(\partial A/\partial V)_T = p \tag{3.21}$$

and

$$-(\partial A/\partial T)_V = S. \tag{3.22}$$

Now, if we apply the cross-differentiation relation (3.8), we obtain an important equation:

$$-(\partial^2 A/\partial V\partial T) = (\partial p/\partial T)_V = (\partial S/\partial V)_T. \tag{3.23}$$

This is one of two thermodynamic equations derived by James Clerk Maxwell in his *Theory of Heat* (Maxwell, 1872). (Maxwell's second equation is considered below.) It illustrates how an important relation – in this case the change of entropy with volume at constant temperature – which is difficult to measure experimentally, may be determined through measurements that are more readily accessible to experiment: the change of pressure with temperature at constant volume. A direct determination of this latter coefficient might also be experimentally difficult, but it can be obtained from two kinds of measurements that are often made. These are determinations of the coefficient of thermal expansion (α) and of the compressibility (κ). They are defined by the relations

$$\alpha = (1/V)\,(\partial V/\partial T)_p, \qquad \kappa = -(1/V)(\partial V/\partial p)_T. \tag{3.24}$$

To see how these coefficients relate to the coefficient $(\partial p/\partial T)_V$ in equation (3.23), consider the volume V as a function of p and T:

$$dV = (\partial V/\partial T)_p\, dT + (\partial V/\partial p)_T\, dp. \qquad (3.25)$$

If we impose on the system the requirement that the volume be held constant ($dV = 0$), we then immediately derive, by the same procedure that led to equation (3.10), the relation

$$(\partial p/\partial T)_V = -\frac{(\partial V/\partial T)_p}{(\partial V/\partial p)_T} = \alpha/\kappa. \qquad (3.26)$$

So the entropy–volume relation in equation (3.23) is obtainable, in straightforward fashion, by measuring the coefficient of thermal expansion and the compressibility of the substances in question.

Actually equation (3.23) was derived, in slightly different form, by the French engineer Clapeyron, with a later modification by Rudolf Clausius, the German physicist who first formulated the concept of entropy. Consider a system composed of a pure liquid in contact with its vapour, at a vapour pressure p. If we were dealing only with a single phase, we could define its state by specifying the values of two controlling variables. Usually we choose these as pressure and temperature. However, in a two-phase system containing only one component, the fact that there is equilibrium between the phases imposes a restriction on the relations between them. In terms of our discussion later in this chapter, we can say that this is the requirement that the chemical potential (μ) of the substance must be the same in both phases: μ (in liquid) = μ (in vapour). Since the system at equilibrium must obey this equation, the value of only one additional variable is needed to determine its state. We can choose a value of either pressure or temperature (within the range of states in which the two phases can coexist): once the value of either one of these variables is fixed, the other is automatically determined. That is why we can define a vapour pressure, p, as a function of T. We consider the system composed of liquid and of the vapour phase above it as being surrounded by a constant temperature bath, which can supply heat to, or withdraw it from, the system, by an infinitesimal change in the temperature of the bath.

If a small quantity of liquid evaporates at constant T, heat being absorbed from the bath to keep the temperature constant, the entropy increase per mole (ΔS) equals the molar heat of vaporization divided by T. Thus, on writing $\Delta S/\Delta V$ instead of $(\partial S/\partial V)_T$, equation (3.23) gives

$$dp/dT = \Delta S/\Delta V = \Delta H/T\,\Delta V. \qquad (3.27)$$

Here we write dp/dT as a total, not a partial, derivative, since p at equilibrium is completely determined if T is specified, and conversely.

Equation (3.27) is commonly known as the Clausius–Clapeyron equation. If the vapour pressure is fairly low, it is a good approximation to treat the vapour as a perfect gas of molar volume TR/p, and to consider the volume of

1 mol of liquid as negligible by comparison with that of 1 mol of vapour. Then $\Delta V = RT/p$ and, after rearrangement, (3.27) becomes

$$(1/p)\,(dp/dT) = d(\ln p)/dT = \Delta H_{vap}/RT^2, \tag{3.28}$$

where ΔH_{vap} is the molar heat of vaporization. For calculation of p as a function of T it is convenient to use the mathematical relation $d(1/T) = -dT/T^2$. when (3.28) assumes the form

$$R\, d(\ln p)/d(1/T) = -\Delta H_{vap}. \tag{3.29}$$

Thus, if we plot $R \ln p$ against $1/T$, the slope of the curve at any point gives the value of $-\Delta H_{vap}$ at that temperature. Conversely, if we know the molar heat of vaporization, and the value of p, at one temperature, we can calculate values of p over a range of temperatures. There will of course be some change of ΔH_{vap} with temperature, corresponding to the change in C_p on vaporization – see equation (3.18) – but this is not large enough to require much correction over a moderate temperature range.

We can apply (3.27) also to the equilibrium between a pure liquid and the corresponding solid phase. In this case the matter of interest is usually the effect of varying pressure on the melting temperature (T_m) of the solid. We therefore write (3.27) in inverted form, setting $T = T_m$, ΔH_m being the enthalpy of fusion:

$$\frac{dT_m}{dp} = \frac{T_m\,\Delta V_m}{\Delta H_m} \quad \text{or} \quad \frac{d(\ln T_m)}{dp} = \frac{\Delta V_m}{\Delta H_m}. \tag{3.27a}$$

Now ΔH_m is always a positive quantity, and for the great majority of substances ΔV_m is also positive. In such cases increase of pressure raises the melting point.

Water, however, is a striking exception; it contracts on melting. At the melting point the molar volume of ice is roughly 19.65 cm^3; that of water is 18.0 cm^3; so $\Delta V = -1.65\ cm^3 = 1.65 \times 10^{-6}\ m^3$. The molar value of $\Delta H = 6000\ J\ mol^{-1}$. If we increase the pressure by 10^5 Pa (approximately by 1 atm) we have (since $T_m = 273$ K), for small finite increments of T_m and p,

$$\Delta T_m = \frac{T_m\,\Delta V_m\,\Delta p}{\Delta H_m} = \frac{273 \times (-1.65 \times 10^{-6}) \times 10^5}{6 \times 10^3} = -0.0075\ \text{K}.$$

In this connection we consider what happens when liquid, solid, and vapour phase of a pure substance are all in equilibrium with each other. Then the chemical potentials of all three phases must be equal (see Section 3.7.):

$$\mu\ (\text{liquid}) = \mu\ (\text{solid}) = \mu\ (\text{vapour}).$$

This gives two equations to determine the state of the system, and there is no room for further variation. If the three phases are to coexist, this is possible only at a single temperature and a single pressure. This is commonly known as a triple point. If the three phases are ordinary ice (ice Ih), liquid water, and water vapour, the temperature is found to be 0.0075 °C and the pressure 600 Pa (0.005 92 atm). If we then raise the pressure to 1 atm, usually by letting in air, the melting point drops to 0 °C, as would be expected from the calculation given above. Now we no longer have the three phases in equilibrium; the system is no longer restricted to a unique pressure and temperature, though the melting temperature remains sharply defined. The analysis developed here represents a special application of the phase rule, which we state in general form later in this chapter.

Proteins and nucleic acids, on heating, commonly undergo reversible transitions between a native state, which predominates below some critical temperature and a denatured state, which predominates above that temperature. Such transitions are not as sharp as the melting of a pure solid; they occur over a limited temperature range, not at a single precisely defined temperature. Nevertheless, the process is regarded as a form of melting, and the effects of pressure on the melting temperature can be dealt with by the same type of analysis that we have described here (see Section 4.10).

Finally we consider the Gibbs energy G, which is often called the Gibbs free energy or simply the free energy. (In the well known text-book by Lewis and Randall (1961) it is called free energy, and denoted by the symbol F.)

The Gibbs energy is defined by the relations

$$G = U + pV - TS = H - TS = A + pV,$$
$$\mathrm{d}G = \mathrm{d}U + p\,\mathrm{d}V + V\,\mathrm{d}p - S\,\mathrm{d}T - T\,\mathrm{d}S = \mathrm{d}H - S\,\mathrm{d}T - T\,\mathrm{d}S. \tag{3.30}$$

Making use of the last expression in (3.30), and inserting the value of dH from equation (3.17), we obtain

$$G = -S\,\mathrm{d}T + V\,\mathrm{d}P = (\partial G/\partial T)_p\,\mathrm{d}T + (\partial G/\partial p)_T\,\mathrm{d}p. \tag{3.31}$$

From the two expressions for dG in (3.31) we obtain

$$(\partial G/\partial T)_p = -S = (G - H)/T; \tag{3.32}$$
$$(\partial G/\partial p)_T = V. \tag{3.33}$$

By cross-differentiation of equations (3.32) and (3.33) we obtain

$$(\partial^2 G/\partial p \partial T) = (\partial/\partial p)\,(\partial G/\partial T)_p = (\partial/\partial T)\,(\partial G/\partial p)_T$$

or

$$-(\partial S/\partial p)_T = (\partial V/\partial T)_p = \alpha V. \tag{3.34}$$

This is the second relation derived by Maxwell. It enables us to calculate, from the coefficient of thermal expansion, the change of entropy with pressure at constant temperature, which would be very difficult to determine by direct measurements.

It follows from the second law of thermodynamics that, in a system at constant pressure and temperature, any spontaneous process must involve a decrease in the Gibbs energy G. A process involving an increase in G cannot occur, unless it is somehow coupled with another process involving a larger decrease, so that the net value of ΔG in the process is negative. The Gibbs energy change for a reaction determines its equilibrium position, as discussed in Chapters 4 and 5. The changes in ΔG for a wide range of processes of interest in biochemistry will be a central concern throughout the rest of this book.

3.5 Applications of the Maxwell equations to water and other liquids

The two Maxwell equations (3.23) and (3.34) relate the change of entropy of a system, with volume and with pressure, to the expansivity α and (for equation (3.23)) to the compressibility κ. The coefficient κ is always positive; that is, a reversible system always contracts when increasing pressure is applied. In the vast majority of systems, α is also positive; when the system is heated, it expands. However, we shall consider one extremely important system, water, for which this statement is not true under some conditions.

If α and κ are both positive, equation (3.23) says that the entropy of a system at constant temperature increases with increasing volume. In terms of the statistical interpretation of entropy, this appears reasonable. Increase of volume should generally increase the number of microscopic states available to the system, and therefore the entropy should increase. Likewise increase of pressure at constant temperature (equation (3.34)), by squeezing the system into a smaller space, might be expected to reduce the number of states available to the system and thereby decrease its entropy. These are very crude qualitative arguments, but all systems in which α is positive behave as these arguments would suggest.

Water below 4 °C is a striking exception. Not only is α slightly negative between 0 and 4 °C, but studies on supercooled water, which have now been carried down to about −40 °C, show that α becomes rapidly more negative as the temperature drops further (Figure 3.1). Table 3.1 gives values of α/κ for water from −20 to +60 °C, showing how striking is the change from negative to positive values of α (and hence of α/κ) as the temperature rises.

We note also that, from (3.34), in the region where α is negative, the entropy of water must *increase* with increasing pressure. We may offer a tentative interpretation in molecular terms. We note that ordinary ice (ice Ih, as distinct from other types of ice formed at high pressure) is much less dense than water (its density is near 0.92 g cm^{-3} as compared to 1 g cm^{-3} for water). Ice has a regular structure, in which each oxygen has four nearest neighbours, arranged tetrahedrally around it, and held by hydrogen bonds. This is a very open structure, with much interstitial space between the atoms. When ice melts, the long range structure breaks down, although the tendency to local tetrahedral order persists in a rather irregular way. The average number of nearest neighbours per oxygen atom becomes greater than in ice, by filling some of the interstitial space; hence cold water is denser than ice. Above 4 °C the normal tendency for the liquid to expand on heating overcomes this

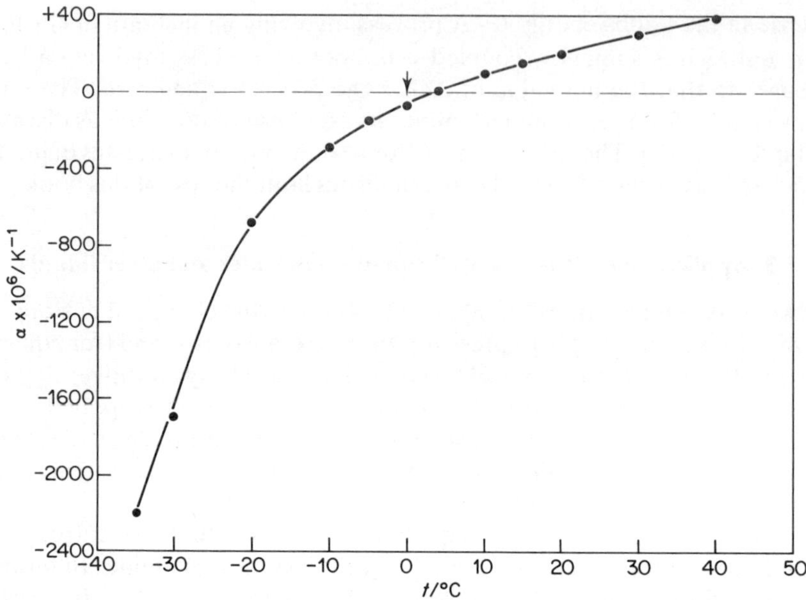

Figure 3.1 The expansivity of water as a function of temperature. (After Angell, Shuppert and Tucker, 1973.)

Table 3.1 Change of entropy of water with volume from the relation $(\partial S/\partial V)_T = (\partial p/\partial T)_V = \alpha/\kappa$

t/°C	$\alpha \times 10^6$/K^{-1}	$\kappa \times 10^{11}$/Pa^{-1}	$10^{-5}(\alpha/\kappa)$/Pa K^{-1}
−20	−680	61.9	−11.0
−10	−295	55.5	− 5.32
0	− 68	51.0	− 1.33
+10	+ 88	47.9	+ 1.84
+20	+207	45.9	+ 4.51
+30	+303	44.8	+ 6.76
+40	+385	44.2	+ 8.71
+50	+458	44.2	+10.36
+60	+523	44.5	+11.75

Data from Kell (1967) and *Handbook of Chemistry and Physics*, 50th edn, p. F–5.

To calculate molar numerical values of $(\partial S/\partial V)_T$, note that in SI nomenclature S is in joules per kelvin per mole and V is in cubic metres. For water $V \simeq 18$ cm^3 = 18×10^{-6} m^3. Thus for 1 mol of water at −20 °C we can write, for a volume increment (ΔV) of 1 cm^3 (= 10^{-6} m^3), $\Delta S = -11 \times 10^5 \times (1 \times 10^{-6}) = -1.1$ J K^{-1} mol^{-1}.

tendency to contraction, and the density diminishes with rising temperature. Below 4 °C the density also diminishes, as we have seen, since α becomes

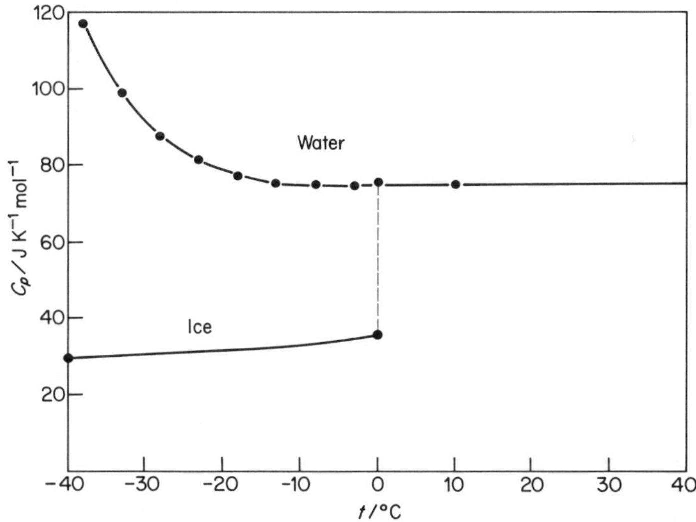

Figure 3.2 The molar heat capacity, C_p, of water and of ice. (After Angell *et al.*, 1973.)

negative. The effect becomes very striking in supercooled water; the density falls from 1.00 g cm^{-3} at 4 °C to 0.973 g cm^{-3} at −35 °C. Qualitatively the most reasonable interpretation seems to be that, as the temperature falls, the water molecules tend to form locally ordered structures, like minute ice crystals on an extremely small scale. These ordered structures, like ice, tend to occupy a larger volume than the less ordered arrangements in ordinary cold water. Increase of pressure tends to inhibit the formation of these ordered structures, since their formation involves expansion of volume. Thus $(\partial S/\partial p)_T$ in this case becomes a positive quantity, since an increase in pressure favours the disordered state, of higher density, and $(\partial S/\partial V)_T$ becomes negative, since an increase in volume favours the more ordered state. This is in striking contrast to the relations found for the vast majority of systems, including water itself above 4 °C.

The heat capacity (C_p) increases markedly as water is supercooled, going from 76 J K^{-1} mol^{-1} at 0 °C to 117 J K^{-1} mol^{-1} at −38 °C. By comparison the value for ice at 38 °C is only 30 J K^{-1} mol^{-1}, and it is only slightly higher at the freezing point (Angell, Schuppert, and Tucker, 1973; Figure 3.2). We have discussed these properties of supercooled water at some length, partly because of their inherent interest and partly because they throw much light on the nature of water, one of the most familiar of all substances and also one of the most extraordinary. For a comprehensive review on supercooled water, see Angell (1983).

It is also illuminating to consider the expansivity and compressibility of water, in comparison with some typical organic liquids, over a range of temperatures at ordinary pressures. Figures 3.3 and 3.4 show this comparison,

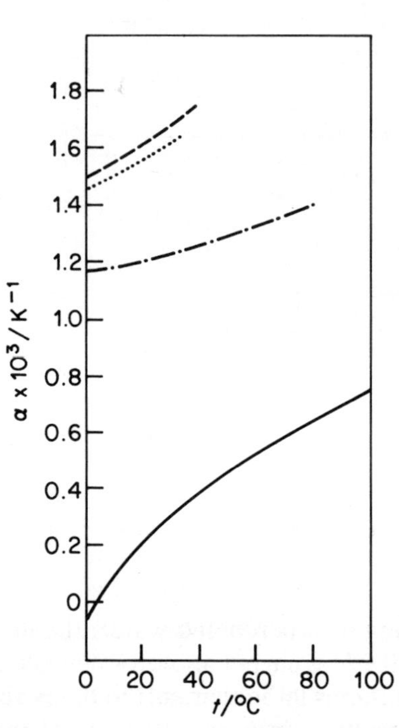

Figure 3.3 Thermal expansivity at atmospheric pressure from 0 to 100 °C for water (—), benzene (-·-), *n*-pentane (····), and diethyl ether (----). (Redrawn from Hvidt, 1978.)

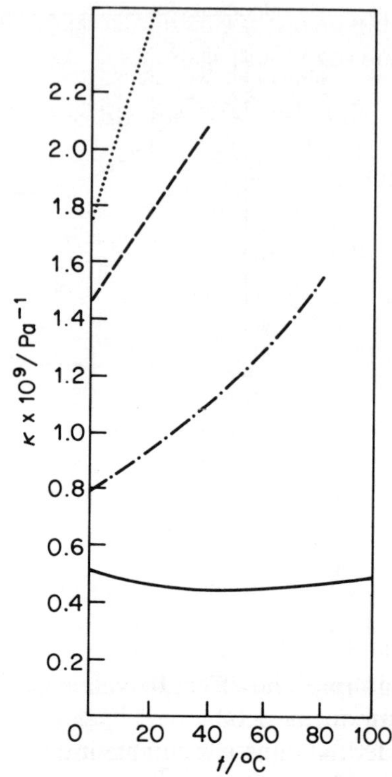

Figure 3.4 Compressibility at atmospheric pressure from 0 to 100 °C for the same molecules as in Figure 3.3. (Redrawn from Hvidt, 1978.)

from 0 to 100 °C, for water, benzene, *n*-pentane, and diethyl ether. It is obvious that all the organic molecules are much more expansible and compressible than water, though $(\partial\alpha/\partial T)_p$ is larger for water than for the others. On the other hand, $(\partial\kappa/\partial T)$ is very small for water, and indeed the compressibility *decreases* slightly with rising temperature to a minimum at 46 °C, and then increases slowly as T increases further. Even so, water is slightly less compressible at 100 °C than at its freezing point. This is among the many remarkable properties of water that still await adequate explanation.

3.6 Systems of two or more components: partial molar quantities

We turn now to systems of two or more components. They contain n_1 moles of component 1, n_2 moles of component 2, etc., and the amounts of any of the various components may be increased or decreased, to vary the composition. Thus we may, for instance, generalize equation (3.31) for the variation in

Gibbs energy by including composition variables as well as T and p; so that for a two-component system it becomes

$$dG = -S\,dT + V\,dp + \mu_1\,dn_1 + \mu_2\,dn_2.$$

Thus

$$dG = \left(\frac{\partial G}{\partial T}\right)_{p,n} dT + \left(\frac{\partial G}{\partial p}\right)_{T,n} dp + \left(\frac{\partial G}{\partial n_1}\right)_{p,T,n_2} dn_1 + \left(\frac{\partial G}{\partial n_2}\right)_{p,T,n_1} dn_2. \tag{3.35}$$

The coefficients μ_1 and μ_2 are known as the chemical potentials of components 1 and 2. In general, for a system of any number of components, the chemical potential of component i is given by

$$\mu_i = (\partial G/\partial n_i)_{p,T,n_j}, \tag{3.36}$$

where the subscript n_j indicates that in adding component i to the system we are holding constant the amounts of all the other components, as well as pressure and temperature. Also μ_i is often termed the partial molar Gibbs energy of component i, since it denotes the increment in G per mole of component i added, at constant composition. It is a differential increment; to determine it accurately we must determine the increment in G for added quantities of the component that are small enough to leave the composition essentially unchanged by the addition.

The study of chemical potentials will be the main theme of this chapter, and of much of the rest of this book. They determine the conditions for chemical equilibrium; for systems not in equilibrium we can say that any component will tend to flow from a region where its potential is higher to one where it is lower, until equilibrium is established. Before pursuing the subject further we must first consider the significance of other partial molar quantities, and the relations between them for the various components.

All the extensive properties of a system – volume, internal energy, enthalpy, entropy, heat capacity, and other properties – also depend on its composition. The partial molar volume of substance i, for instance, denotes the increment V_i:

$$V_i = (\partial V/\partial n_i)_{p,T,n_j} \tag{3.37}$$

denotes the volume increment per mole of component i added, in a system of given composition. We choose the relations among the volumes of the components to illustrate some general properties of partial molar quantities.

The volume increment, in a system of m components at constant T and p, can be written as follows:

$$\mathrm{d}V = \left(\frac{\partial V}{\partial n_1}\right)_{n_j} \mathrm{d}n_1 + \left(\frac{\partial V}{\partial n_2}\right)_{n_j} \mathrm{d}n_2 + \ldots + \left(\frac{\partial V}{\partial n_m}\right)_{n_j} \mathrm{d}n_m$$

$$= \sum_{i=1}^{m} \left(\frac{\partial V}{\partial n_i}\right)_{n_j} \mathrm{d}n_i = \sum_{i=1}^{m} V_i \mathrm{d}n_i. \tag{3.38}$$

Now we can in effect integrate this equation by adding small (or in the limit infinitesimal) increments of all the components, in amounts proportional to the initial composition, and continue to make the system conceptually as large as we please, keeping the composition unchanged throughout the process. Thus all the V_i, the partial molar volumes, remain unchanged throughout the enlargement of the system and can be treated as constants during the integration. The amount of component i is n_i^0 at the start, and $\alpha n_i^0 = n_i$ at the end, where α is a very large number which is the same for all components. Thus the volume V of the whole system at the end of the process is given by

$$V = \sum_{i=1}^{m} V_i \int_{n_i^0}^{\alpha n_i^0} \mathrm{d}n_i = \sum_{i=1}^{m} V_i\, n_i = \sum_{i=1}^{m} n_i \frac{\partial V}{\partial n_i}\,. \tag{3.39}$$

We can make n_i° negligibly small compared to αn_i°, so that in effect we are integrating $\mathrm{d}n_i$ from zero up to the final value of n_i.

This reasoning is quite general, and exactly analogous equations can be written for any extensive property of a thermodynamic system, i.e. for U, H, S, G, or C_p, for example.

Equation (3.39) can also be derived by another route, which emphasizes a general relation for homogeneous functions – an important class of functions in thermodynamics and elsewhere. A function $f(x,y,z, \ldots)$ is homogeneous of order n if, on multiplying each of the independent variables by an arbitrary factor λ, the function itself is multiplied by λ^n:

$$f(\lambda x, \lambda y, \lambda z) = \lambda^n f(x,y,z).$$

For instance let

$$f(x,y) = ax^2 + bzy + cz^2;$$

then

$$f(\lambda x, \lambda y) = a\lambda^2 x^2 + b\lambda^2 xyz + c\lambda^2 z^2 = \lambda^2 f(x,y).$$

Thus this function is homogeneous of second order.

Leonhard Euler, the great eighteenth century mathematician, derived a fundamental theorem concerning such functions. We state Euler's theorem for a function of two variables; the reasoning can be immediately extended to any number of variables. The theorem states that, if $f(x,y)$is homogeneous of degree n, then

$$x\left(\frac{\partial f}{\partial x}\right)_y + y\left(\frac{\partial f}{\partial y}\right)_x = nf(x,y). \tag{3.40}$$

To prove it we may proceed as follows. Set $x^* = \lambda x$ and $y^* = \lambda y$. Then, since $f(x,y)$ is homogeneous,

$$f^* = f(x^*,y^*) = f(\lambda x, \lambda y) = \lambda^n f(x,y).$$

The total differential of f^* is

$$\mathrm{d}f^* = \frac{\partial f^*}{\partial x^*}\,\mathrm{d}x^* + \frac{\partial f^*}{\partial y^*}\,\mathrm{d}y^*.$$

So

$$\frac{\mathrm{d}f^*}{\mathrm{d}\lambda} = \frac{\partial f^*}{\partial x^*}\frac{\mathrm{d}x^*}{\mathrm{d}\lambda} + \frac{\partial f^*}{\partial y^*}\frac{\mathrm{d}y^*}{\mathrm{d}\lambda} = x\frac{\partial f^*}{\partial x^*} + y\frac{\partial f^*}{\partial y^*}\cdot \tag{3.41}$$

Also

$$\frac{\mathrm{d}f^*}{\mathrm{d}\lambda} = \frac{\mathrm{d}f(x^*,y^*)}{\partial\lambda} = \frac{\mathrm{d}}{\mathrm{d}\lambda}[\lambda^n f(x,y)] = n\lambda^{n-1}f(x,y). \tag{3.42}$$

On equating the right-hand sides of equations (3.41) and (3.42) we obtain

$$\frac{\partial f^*}{\partial x^*}x + \frac{\partial f^*}{\partial y^*}y = n\lambda^{n-1}f(x,y). \tag{3.43}$$

However, since λ is an arbitrary parameter we can set $\lambda = 1$, when (3.43) becomes identical with (3.40), which is Euler's theorem.

The volume V of a thermodynamic system, at constant p and T, is a homogeneous function of the amounts of the various components, n_1, n_2, ..., n_m. It is homogeneous of the first order, since if we multiply all the n_i by a given factor, V is multiplied by the same factor. Thus for $V(n_1, n_2, ..., n_\mathrm{m})$, by Euler's theorem, we again obtain equation (3.39):

$$V = n_1\frac{\partial V}{\partial n_1} + n_2\frac{\partial V}{\partial n_2} + \ldots + n_m\frac{\partial V}{\partial n_m} = \sum_{i=1}^{m} n_iV_i. \tag{3.39}$$

Since V is a function of the n_i and also of the V_i, we can differentiate it to obtain the following general form:

$$\mathrm{d}V = \sum_{i=1}^{m} V_i\mathrm{d}n_i + \sum_{i=1}^{m} n_i\mathrm{d}V_i. \tag{3.44}$$

However, we know from equation (3.38) that $\mathrm{d}V = \sum V_i\,\mathrm{d}n_i$. Hence from (3.44) it follows that

$$\sum_{i=1}^{m} n_i \mathrm{d}V_i = 0. \tag{3.45}$$

It is easiest to see the meaning of (3.45) if we consider a two-component system at constant T and p. Then (3.45) becomes

$$n_1 \, \mathrm{d}V_1 + n_2 \, \mathrm{d}V_2 = 0.$$

Suppose that we add a small increment of component 1 to the system, keeping n_2 constant. Then

$$n_1 \left(\frac{\partial V_1}{\partial n_1}\right)_{n_2} + n_2 \left(\frac{\partial V_2}{\partial n_1}\right)_{n_2} = 0. \tag{3.46}$$

On rearranging terms we find

$$\frac{(\partial V_2/\partial n_1)_{n_2}}{(\partial V_1/\partial n_1)_{n_2}} = -n_1/n_2. \tag{3.47}$$

So, if adding component 1 increases V_1, we see from (3.46) that this addition must correspondingly decrease V_2, and the ratio of the two increments is fixed by (3.47). If $n_1 = n_2$, the two differential coefficients on the left of (3.46) must be equal, but of opposite sign. There are some binary systems for which V_1 and V_2 are essentially independent of composition, and equal to the molar volumes of the pure components. A mixture of benzene and toluene fits such a relation closely; their similarity is so great that a benzene molecule 'sees' a toluene molecule in its neighbourhood almost as if it were another benzene molecule. The solution approximates closely to what we shall later discuss as an ideal solution.

In equations (3.45), (3.46), and (3.47) we can replace the volume V by any other extensive property, and the partial molar volume (V_i) of component i by the partial derivative of such an extensive property with respect to moles (or some other units) of component i. This only requires that the equations for the extensive property be first order and homogeneous. The relations for the derivatives of the Gibbs energy (chemical potentials) that correspond to the equations above are particularly important (see Section 3.10).

Figure 3.5 shows the apparent molar volumes of propan-2-ol in water in a two-component solution over the whole range of composition. Obviously these solutions are far from ideal. For ethanol or propan-2-ol in very dilute solution in water, V_2 is smaller, by about 5 cm^3 mol^{-1}, than V_2^0 for pure ethanol, which is close to 61 cm^3 mol^{-1}. In dilute solutions, up to a mole fraction of about 0.05, V_2 decreases still further, then passes through a minimum and increases at higher x_2 values. This behaviour is characteristic of substances containing non-polar groups (such as the propyl group in propanol) when dissolved in water at high dilution; that is, V_2 in water at x_2 near 0 is always

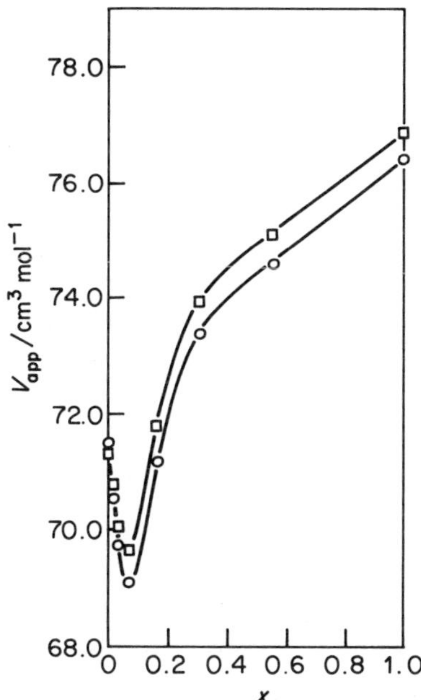

Figure 3.5 The apparent molar volume (V_{app}) of propan-2-ol as a function of its mole fraction x: ○, at 20 °C; □ at 25 °C. (Redrawn from Bruun and Hvidt, 1977.)

lower than V_2^0 for the pure organic liquid. This is one aspect of what are termed hydrophobic interactions between water and compounds containing non-polar groups. In some cases the decrease in V_2 is very marked indeed. For instance Kliman (1969) – cited by Kauzmann (1974) – has found, for octan-2-one, $CH_3(CH_2)_5COCH_3$, that V_2 in water at low concentration at 25 °C is nearly 35 cm^3 mol^{-1} lower than for pure octan-2-one.

No substance, of course, can occupy a negative volume; but negative values of partial molar volumes can and do occur. Thus if we add magnesium sulphate to water, at low concentration, below 0.05 M, the volume of the liquid actually shrinks as a result of the addition. As zero concentration is approached the value of V_2 approaches -8 cm^3 mol^{-1}. At concentrations above about 0.07 M, V_2 becomes positive, and increases still further at higher concentrations.

From the point of view of pure thermodynamics, we do not need to ask why V_2 is negative for a dilute $MgSO_4$ solution; accurate density measurements on solvent and solution show this to be a fact, and we can use the resulting value with assurance in further calculations. However, we may remark on the interpretation in molecular terms, since the general idea is illuminating. Both Mg^{2+} and SO_4^{2-} are bivalent ions, and Mg^{2+} in particular is a very small ion. Around these ions in water there is an intense electric field gradient, as much as 160 kV cm^{-1} in the immediate neighbourhood of the ion, which falls off rapidly with distance. In such an intense and inhomogeneous field the water molecules around the ions, being dipoles, orient themselves around the ions, and are packed close to them, occupying considerably less space than in bulk water. This is particularly true of small cations such as Mg^{2+}. In this case the packing effect, known as electrostriction, is so large that it more than compensates for the volume increment due to the bare ions themselves. All ions produce some electrostriction in aqueous solution, but in most cases the increase in volume due to the ions themselves is greater than the decrease due to electrostriction, and the partial molar volumes are positive, except in a few cases such as dilute $MgSO_4$ solution.

3.6.1 Apparent molar volumes

In determining partial molar volumes for solutes in water at high dilution, it is often useful to express experimental data in terms of a related quantity, the apparent molar volume, often written V_2^ϕ. This is simply defined as the volume of the solution (V), minus the volume of the pure solvent that it contains, divided by the number of moles of solute (component 2).

$$V_2^\phi = (V - n_1 V_1^0)/n_2. \tag{3.48}$$

For an aqueous solution V_1^0 is the volume of 1 mol of water, at the temperature and pressure in question. The value of V_2^ϕ is readily calculated from density measurements of solvent and solution. It is not in itself a quantity of direct thermodynamic significance, but in dilute solution V_2^ϕ is a close approximation to the true partial volume V_2, and the two become equal at infinite dilution. Therefore a determination of V_2^ϕ at several concentrations of solute, followed by an extrapolation of the data to infinite dilution, gives the limiting value of V_2 as n_2 approaches zero. For practical purposes the value of V_2^ϕ at concentrations below 0.05–0.10 M is often quite sufficient.

Long ago Traube (1899) pointed out that the apparent molar volumes of organic compounds in water could be calculated with fair accuracy, by summing volume contributions from the constituent groups in the molecule and then adding a roughly constant term which he called the covolume. Some of Traube's constituent volumes for atoms were (in cubic centimetres per mole): C 9.9; H 3.1; N 1.5; O (hydroxyl) 0.4; O (carbonyl) 5.5.

To the sum of those values for a particular compound, it was necesary to add a 'covolume' term, nearly independent of the size of the molecule. For an organic substance in dilute solution in water, Traube estimated the covolume as approximately 13 $cm^3\ mol^{-1}$; for the same substance as a pure organic liquid, the covolume was substantially larger, close to 25 $cm^3\ mol^{-1}$.

There have been other proposals for calculating partial or apparent molar volumes for summation of terms for constituent groups. In Table 3.2 we have listed some values, taken from a recent paper by Perron and Desnoyers

Table 3.2 Calculation of partial molar volumes of solutes in aqueous solution from contributions of constituent groups

Group	V_2^0/cm^3 mol^{-1} at 25 °C	
	Aliphatic	Aromatic
—H	10.7	10.1
—CH_3	26.7	23.2
—OH	12.0	12.6
—NH_2	15.1	14.9
—CH_2—	16.0	14.6
—O—	4.1	6.9
—COOH	25.5	25.7
—CONH—	20.0	–
—CHO	–	23.0
Benzene	–	83.2

Values are from Perron and Desnoyers (1979), except that for —CONH—, which is derived from Traube (1899) as applied by Cohn and Edsall (1943, Chapter 7). For ions and dipolar ions in water an electrostriction term must be subtracted from the sum of the constituent volumes to obtain correct results; see text and Table 3.3. For monosubstituted benzene derivatives, take the value for benzene, subtract the value for —H, and add the value for the substituent group. Thus for toluene $V_2 = 83.2 - 10.1 + 23.2 = 96.3$ cm^3 mol^{-1}.

(1979). They have distinguished between groups in aliphatic and aromatic compounds, and their mode of calculation requires no covolume term.

The electrostriction effect, due to ionic groups in aqueous solution, has an important influence on partial molar volumes. When two neutral molecules in water solution give rise to ions by proton transfer between them, there is almost always a marked contraction in volume. Thus for water itself, and for acetic acid,

$$2H_2O \rightleftharpoons H_3O^+ + OH^-, \ \Delta V = -21 \text{ cm}^3 \text{ mol}^{-1};$$

$$CH_3COOH + H_2O \rightleftharpoons H_3O^+ + CH_3COO^-, \ \Delta V = -10.3 \text{ cm}^3 \text{ mol}^{-1}.$$

Molecules such as the amino acids, which contain ionic groups even at their isoelectric points, also exert electrostriction on the surrounding water molecules. This is readily seen when we compare their partial molar volumes with the values for isomeric substances that do not contain such ionic groups. The data of Table 3.3 illustrate this. The value for amino acids is always less that for their uncharged isomers, and the difference gives the estimated electrostriction, E. For isomers with different numbers of atoms separating the charged groups – α- and β-alanine, or α- and ε-aminocaproic acid for instance – it is seen that the electrostriction increases with the separation

Table 3.3 Some partial molar volumes (V_2) of amino acids and related isomeric substances at high dilution in water

Substance	Formula	$V_2/cm^3\ mol^{-1}$	$E/cm^3\ mol^{-1}$
Glycolamide	$CH_2OHCONH_2$	56.2	
Glycine	$^+H_3NCH_2COO^-$	43.5	12.7
Lactamide	$CH_3CHOHCONH_2$	73.8	
α-Alanine	$^+H_3NCH(CH_3)COO^-$	60.6	13.2
β-Alanine	$^+H_3NCH_2CH_2COO^-$	58.9	14.9
Methylhydantoic acid	$H_2NCONHCH(CH_3)COOH$	94.2	
Glycylglycine	$^+H_3NCH_2CONHCH_2COO^-$	77.2	17.0
α-Aminocaproic acid	$CH_3(CH_2)_3CH(NH_3^+)COO^-$	108.4	12.1
ε-Aminocaproic acid	$^+H_3N(CH_2)_5COO^-$	104.9	15.7

Data from Cohn and Edsall (1943), pp. 158–159.

The electrostriction, E, for glycine, α- and β-alanine, and glycylglycine is calculated by subtracting V_2 for the amino acid from V_2 for its uncharged isomer. For the two aminocaproic acids it is calculated by adding the constituent group volumes (see Table 3.2).

between the charges. Thus the partial (and apparent) molar volumes of ions and molecules in aqueous solution reflect both the structure of the solute and the interactions of the solute molecules with the surrounding medium.

We have discussed these relations in terms of partial and apparent molar volumes, but the same general relations hold for any other choice for defining the composition of a solution. Often it is useful to consider the partial specific volume of a component, defined by the relation

$$v_i = \left(\frac{\partial V}{\partial w_i}\right)_{T,p,w_j}.$$

Here v_i is the differential volume increment per gram of substance i added, and w_i is its concentration in grams per unit volume. In dealing with a macromolecule of unknown molecular weight, the determination of its partial specific volume is an essential step for determining its molecular weight in the ultracentrifuge.

For a solution of m components, relations of the same form as equation (3.39) hold; for instance, for the total volume V,

$$V = w_1 v_1 + w_2 v_2 + \dots + w_m v_m.$$

3.7 Chemical potentials: definitions

We have already introduced the chemical potential of a component in equations (3.35) and (3.36). It provides a quantitative measure of what G. N. Lewis called the 'escaping tendency' of the component; that is, in a system not at equilibrium, any component tends to flow from regions in the system where its escaping tendency (or potential) is high to those where it is lower. The concept of chemical potential bears a certain analogy to that of temperature. Just as a system in thermal equilibrium must be at constant temperature throughout, so for a system at constant pressure and temperature, in equilibrium, the chemical potential of each component must be the same throughout the system, in any phase in which it is present. Thus, like temperature, chemical potential is an intensive quantity; the potential of any component, by the definition given below, does not depend on the size of the system, but only on its pressure, temperature, and composition. For a dilute gas the chemical potential varies as the logarithm of the concentration, and the same is approximately true for many substances in solution.

We repeat the definition of chemical potential as the increment in Gibbs energy† that accompanies the addition to the system of a unit amount of that component, at constant temperature, pressure, and composition:

†This is the only definition of chemical potential that we shall need in practice. However, the potential can be defined in other, equivalent ways. For instance, equation (3.20) for the variation in the Helmholtz energy, A, in a system of variable composition is extended to become

$$\mu_i = \left(\frac{\partial G}{\partial n_i}\right)_{p,T,n_j} \qquad (3.50)$$

The subscript n_j in (3.50) indicates that the amounts of all components, other than i, are held constant during the addition. Moreover μ is a differential increment; that is, the amounts of all components present, including i, must be very large in relation to the increment dn_i, so that the addition may be regarded as occurring at constant composition. We are free to choose the units that define the amounts of the components present; they can be grams or kilograms or moles, or any other unit that serves our convenience. Ordinarily, however, we shall express amounts of substance in moles, and the molar chemical potential of a component will be taken as the differential increment in Gibbs energy, per mole of that component added to the system, at constant p, T, and composition. We shall employ Gibbs' symbol μ for potential to denote such molar quantities, except in unusual cases, in which the choice of a different unit of quantity will be explicitly stated in each case.

The molar chemical potential of a component is often called its partial molar Gibbs energy (or in earlier texts the partial molar free energy). The basis for using this term is apparent from equation (3.50). However, we shall follow Gibbs in using the term potential and the symbol μ for this quantity.

3.8 Chemical potentials in ideal gases

The simplest example of a chemical potential is that of a single pure substance in the form of an ideal gas. In this case the potential of the substance is identical with its molar Gibbs energy at the given pressure and temperature. We designate it as component 1, and consider the dependence of its potential on the pressure of the gas. This follows from the fundamental equation relating G and p, in a system containing n_1 moles of the gas:

$$\begin{aligned} dA &= -SdT - pdV + \sum_{i=1}^{m} \left(\frac{\partial A}{\partial n_i}\right)_{T,V,n_j} dn_i \\ &= -SdT - pdV + \sum_{i=1}^{m} \mu_i dn_i. \end{aligned}$$

Thus the chemical potential μ_i is given by

$$\mu_i = (\partial A/\partial n_i)_{T,V,n_j}.$$

In studies of gases, it is experimentally feasible to hold the temperature and volume constant while varying the composition, and A can therefore be a useful function in gaseous systems. Liquids and solids, however, have such low compressibilities that we cannot add a component to the system, at constant T and V, except by raising the pressure to quite abnormal levels. Hence G, which is the appropriate function for work at constant temperature and pressure, is the basis for determining chemical potentials in all biochemical work.

By similar extensions of equations (3.13) and (3.17) we can also derive the relations

$$\mu_i = (\partial U/\partial n_i)_{S,V,n_j} = (\partial H/\partial n_i)_{S,P,n_j}.$$

These definitions, however, are not useful as a basis for experimental work, for it is not practicable to add a component to a system while keeping the entropy constant.

$$\frac{1}{n_1}\left(\frac{\partial G}{\partial p}\right)_T = \left(\frac{\partial \mu_1}{\partial p}\right)_T = V_1 = \frac{RT}{p}\,.$$

It is important to consider how the potential of the gas varies with its pressure, and therefore with its concentration. If we change the pressure, at constant temperature, from p_A to p_B, the change of potential is

$$(\mu_i)_B - (\mu_1)_A = \int_{p_A}^{p_B} V\,dp = \int RT\,d(\ln p)$$

$$= RT\,\ln\left(\frac{p_B}{p_A}\right) = RT\,\ln\left(\frac{c_B}{c_A}\right). \qquad (3.52)$$

Since $p = RTc$, where c is the concentration of the gas in moles per litre, the last equality in (3.52) follows directly. This linear relation between the change in chemical potential of an ideal gas, and the change in the logarithm of its concentration, does not hold strictly for real systems. However, it is good enough, as a first approximation, to describe gases at low pressures and most solutes in dilute liquid solutions. For most work in biochemistry, the approximation is adequate.

The change in chemical potential, when a substance passes from one state to another, is directly measurable. To obtain a numerical value of the potential in any one state, however, we must refer it to some standard state, which we are free to choose. The standard state usually chosen for gases is a pressure of one standard atmosphere (1 atm = 101.325 kPa) at $T = 273.15$ K. If, for example, in equation (3.52), $p_A = 1$ atm and $p_B = 0.1$ atm, the potential of the gas at p_B would be

$$\mu_1 - \mu_1^\circ = RT\ln(p_1/p_1^\circ) = RT\ln(0.1/1) = -5.72 \text{ kJ mol}^{-1}$$
$$= -1.365 \text{ kcal mol}^{-1}.$$

We might, however, choose our standard state for the gas as a concentration of 1 mol dm^{-3}. If $p = 0.1$ atm at 25 °C, this corresponds for an ideal gas to $c = 2.45$ mol dm^{-3}. In this case we have

$$\mu_1 - \mu_1^\circ = RT\ln(c_1/c_1^\circ) = RT\ln(2.45/1) = +2.23 \text{ kJ mol}^{-1}$$
$$= +\,0.533 \text{ kcal mol}^{-1}.$$

The relations between standard chemical potential, standard Gibbs energy and equilibrium constant are discussed further in Section 4.2.

We emphasize these differences in standard state to remind the reader that the numerical value of a chemical potential has meaning only in reference to a given standard state, which must always be clearly specified. In the reporting of data in the scientific literature, authors have often failed to state clearly their definition of the standard state that they had chosen; so we emphasize this point. On the other hand, of course, the *difference* in chemical potential accompanying the transfer of a substance from one actual state to another – as in the change of gas pressure from p_A to p_B in equation (3.52) above – is independent of the standard state, since the latter cancels out in taking the difference. This discussion relates to ideal gases, which means that it is applicable only to pressures of 1 atm or less. If systems at high gas pressures are considered, the difference between pressure and a term called fugacity must be taken into account. For a discussion of this special problem the reader is referred to Lewis and Randall (1961, Chapters 14 and 16) and Klotz and Rosenberg (1972).

In a mixture of gases, again assumed to be ideal, we take the partial pressure (p_i) of component i as equal to the total pressure (p), multiplied by the mole fraction (x_i) of component i in the mixture:

$$p_i = px_i = p_i n_i / \sum n_i. \tag{3.53}$$

Here $\sum n_i$ denotes the sum $n_1 + n_2 + \ldots + n_r$, taken over all the r components of the system. The same definition of mole fraction, of course, applies to a component of a liquid or solid phase.

3.9 Chemical potential and activity: activity coefficients

Since changes in chemical potential are often approximately linear in the logarithm of the concentration of the component in question, it is useful to employ a function called the activity, which likewise bears a logarithmic relation to the potential. By definition the activity a_i of component i is related to its potential by the equation

$$\mu_i - \mu_i^\circ = RT \ln a_i. \tag{3.54}$$

We note that for an ideal gas $a_i = p_i/p_i^\circ = c_i/c_i^\circ$, where p_i° is the pressure (or partial pressure) of the gas in the standard state, and c_i° is the concentration in this state. If the gas is in equilibrium with a liquid phase containing component i, the potential of that component must be the same in the two phases. The same will be true of the activity, *provided that we have chosen the standard state to be the same for both phases.* It is often but not always convenient to choose the same standard state for both. In any case the ratio of the activities in the two phases, at equilibrium, will be a constant, the constant being unity if the same standard state is chosen for both.

Whatever standard state is chosen (at a given temperature) the activity is always unity in that state *by definition*, and the activity in any other state is relative to that in the standard state. Thus the activity is a pure number, and is dimensionless. This of course is to be expected from the form of equation

(3.54): any logarithm must be the log of a pure number.† It is meaningless to speak of the logarithm of a pressure, for instance. It is only the numerical value of the pressure that can be used for this purpose.

Since the activity of a substance in solution is often proportional to its concentration, it is frequently convenient to express data in terms of a ratio of activity to concentration, called an activity coefficient. There are several ways of expressing concentrations; all are closely related, but they must be distinguished. In biochemical work, concentrations are commonly expressed in moles per unit volume, generally moles per cubic decametre (moles per litre). If c_i is the volume concentration of component i, then the activity coefficient of component i, γ_i, is given by

$$\gamma_i = a_i/c_i. \qquad (3.55)$$

For solutes at very low concentration in water, we generally fix the value of γ_i by the convention that $a_i = c_i$ and $\gamma_i = 1$, in the limit at infinite dilution of solute.‡ The standard state then is an 'ideal' concentration of 1 mol dm^{-3}.

Volume concentrations represent the most practically convenient units in biochemistry. They also have advantages from a theoretical point of view, as Ben-Naim (1978) has pointed out, in terms of an argument based on statistical mechanics, which we cannot discuss here. They have one obvious disadvantage: they change if the temperature varies, since the volume of the solution changes.

For this among other reasons, many chemists prefer to report data in terms of masses (or moles) of components. The total mass of a system that is closed to transfer of matter must of course remain constant,§ and the masses of the individual components are also constant, unless change of temperature produces a shift in a chemical equilibrium (see Chapter 4). A widely used convention is to express masses as moles of solute per unit mass (usually 1 kg) of solvent. The choice of one component as the solvent is sometimes arbitrary, but in the predominantly aqueous systems that are usually of major concern to biochemists water is generally denoted as the solvent. This concentration unit, commonly denoted the molality, m_i, for component i is given by

$$m_i = n_i/\text{kg}\,H_2O = \text{moles of component } i/55.51 \text{ mol } H_2O. \qquad (3.56)$$

†The activity as defined here is thus a relative activity, as defined by G. N. Lewis. A different but related concept, that of absolute activity, is employed by some authors, notably by E. A. Guggenheim (1967) in his *Thermodynamics*. We shall not make use of absolute activities here.

‡Since γ_i is a pure number and a_i is dimensionless, c_i in (3.55) must be dimensionless also. This point has puzzled many students, but the matter is clarified if we recognize that c_i is actually a relative concentration, c_i/c_i°, where c_i° is the concentration in the standard state. Since this is usually 1 mol dm^{-3}, we implicitly set $c_i^\circ = 1$, and omit it for practical purposes in writing the equations.

§For all practical purposes in chemistry we can ignore the change of mass that occurs when energy is evolved or absorbed; this is immeasurably small. Consider a process with $\Delta U = 10^6$J, a large amount for a chemical reaction. From Einstein's equation for the transformation of mass into energy, $\Delta m = \Delta U/c^2$. Here c, the velocity of light $= 3 \times 10^8$ m s^{-1}. Hence $\Delta m = 10^6/(9 \times 10^{16}) \simeq 10^{-11}$ kg $\simeq 10^{-8}$ g. Even with an extremely delicate balance, it would be impossible to measure such a *change* in mass.

The corresponding activity coefficient is

$$\gamma_i = a_i/m_i. \tag{3.57}$$

We use the same symbol γ_i to denote the activity coefficient for this system of concentration units. If it is necessary to discriminate the value of γ_i for molar units, we can denote it as $(\gamma_i)_m$.

For some important purposes, the most rational choice of a concentration unit is the mole fraction, which we have already defined. The activity coefficient on this basis is

$$\gamma_i = a_i/x_i. \tag{3.58}$$

If we need to discriminate γ_i, expressed on this basis, from the γ values for other concentration units, we can write it with a subscript x, as $(\gamma_i)_x$. Usually, however, if the context is clear, we can omit such subscripts.

If we are dealing with a pure substance, its mole fraction is of course unity, and we take the activity as equal to unity in this state, so $\gamma_i = 1$ in this state. In a solution, when one component is present in much higher concentration than any of the others, it is usually termed the solvent, and its activity is taken as unity in the pure state. In biochemical systems it is therefore common to treat water as the solvent, and to set $a = 1$ for pure water. Then if, for instance, the mole fraction of water is 0.95, the activity of water would be 0.95 if the solution could be regarded as ideal. In this case the vapour pressure of the water would also be 0.95 times that of pure water ($p°$) at that temperature, and γ_{water} would be 1. If the vapour pressure were higher or lower than 0.95 $p°$, γ would be correspondingly higher or lower than unity. The further discussion will illustrate this point in detail.

3.10 Linked and reciprocal relations among chemical potentials of different components: the Gibbs–Duhem equation and the Bjerrum equation

The variation of Gibbs energy in a system defined in terms of pressure, temperature, and composition is

$$\begin{aligned} \mathrm{d}G &= -S\,\mathrm{d}T + V\,\mathrm{d}p + \mu_1\,\mathrm{d}n_1 + \mu_2\,\mathrm{d}n_2 + \ldots \\ &= -S\mathrm{d}T + V\mathrm{d}p + \sum_{i=1}^{m} \mu_i \mathrm{d}n_i . \end{aligned} \tag{3.59}$$

At constant T and p, this becomes

$$\mathrm{d}G = \sum \mu_i\,\mathrm{d}n_i. \tag{3.59a}$$

By the same reasoning, from Euler's theorem, that we have employed in relating the total volume of a system to its composition and the partial volumes of its components (equations (3.39)–(3.43) inclusive), we can integrate equation (3.59a) to give

$$G - G° = \sum \mu_i n_i, \tag{3.59b}$$

where G° denotes a reference state of Gibbs energy for the system taken as a whole. On differentiating (3.59b) we obtain

$$dG = \sum_{i=1}^{m} \mu_i \, dn_i + \sum_{i=1}^{m} n_i \, d\mu_i. \tag{3.59c}$$

On subtracting (3.59a) from (3.59c) we obtain

$$\sum_{i=1}^{m} n_i \, d\mu_i = 0. \tag{3.59d}$$

This important relation is exactly analogous to equation (3.45) for partial volumes. It was first derived by Gibbs. Its use was particularly emphasized by the French physical chemist Pierre Duhem, and it is commonly called the Gibbs–Duhem equation. For a two-component system it becomes

$$n_1 \, d\mu_1 + n_2 \, d\mu_2 = 0 \tag{3.59e}$$

or, in terms of activities (see equation 3.54),

$$n_1 \, d(\ln a_1) + n_2 \, d(\ln a_2) = 0. \tag{3.59f}$$

Thus, if we add an increment dn_1 of component 1,

$$n_1 \, d(\ln a_1)/dn_1 = -n_2 \, d(\ln a_2)/dn_1. \tag{3.59g}$$

We may also write these relations in terms of mole fractions, by dividing (3.59e) or (3.59f) by $n_1 + n_2$. Since

$$x_1 = n_1/(n_1 + n_2) \quad \text{and} \quad x_2 = 1 - x_1;$$

hence $dx_2 = -dx_1$. Then

$$x_1 \, d(\ln a_1)/dx_1 + x_2 \, d(\ln a_2)/dx_1 = 0.$$

Hence

$$x_1 \, d(\ln a_1)/dx_1 = -x_2 \, d(\ln a_2)/dx_1 = x_2 \, d(\ln a_2)/dx_2$$

or

$$d(\ln a_1)/d(\ln x_1) = d(\ln a_2)/d(\ln x_2). \tag{3.59h}$$

Thus, if the activity of one component increases, the activity of the other must decrease. Moreover, these equations specify quantitatively the relations between the changes in the activities of the two components. We consider shortly their application to a specific system.

For systems of more than two components, the Gibbs–Duhem equation does not lead to such simple relations. We can say, however, that if (for instance) $d\mu_1$ is positive, then the sum of the $d\mu$ values for all the other components must be negative, and the sum of all the other $n_i \, d\mu_i$ terms must be equal to $\mu_1 \, dn_1$.

There is, however, another equation that relates the variations in chemical potential between any two components in a multicomponent system. It follows simply from the fact that G, like μ, H, or S, is defined by the state of the system, independently of the path to that state. Thus dG is an exact differential. On differentiating G successively with respect to the amounts of any two components, denoted as i and k, we obtain

$$\left(\frac{\partial \mu_i}{\partial n_k}\right)_{p,\ T,\ n_j} = \left(\frac{\partial^2 G}{\partial n_i \partial n_k}\right) = \left(\frac{\partial \mu_k}{\partial n_i}\right)_{p,\ T,\ n_j} . \tag{3.60}$$

The subscript n_j denotes that the amounts of all components, except the one being added to the system, are held constant. This relation follows directly from the fundamental equations of Gibbs. It appears to have been first explicitly used by Niels Bjerrum (1923) and Bjerrum's name is therefore often attached to it. We may illustrate its use by considering a solution of haemoglobin in equilibrium with oxygen and carbon dioxide, each gas being present at a specified partial pressure, and therefore at a specified potential. If we increase the amount of CO_2 in the system, it has been known since the work of Bohr, Hasselbalch, and Krogh (1904) that oxygen will be driven off from the liquid into the gas phase; that is, in a closed system, the potential of oxygen will increase. In this case the Bjerrum equation becomes

$$\left(\frac{\partial \mu_{O_2}}{\partial n_{CO_2}}\right)_{p,\ T,\ O_2} = \left(\frac{\partial \mu_{CO_2}}{\partial n_{O_2}}\right)_{p, T,\ CO_2} . \tag{3.61}$$

Thus we see immediately that, if adding CO_2 increases the potential (and therefore the partial pressure) of oxygen, then it follows inevitably that addition of oxygen must increase the partial pressure of CO_2; it will drive off CO_2 from the solution into the gas phase. Bohr did not realize that such a reciprocal relation must exist; he sought for it experimentally and failed to find it. It was only in 1914 that J. S. Haldane and his collaborators (Christiansen, Douglas, and Haldane, 1914) demonstrated it experimentally, as one aspect of what is now generally called the Bohr effect.

Equations (3.60) and (3.61) exemplify what is commonly called a linkage relation between the potentials, or activities, of different components, in systems of any degree of complexity. From it, and related expressions, we can develop a very large number of other linkage relations between the activities of the different components in a system (Wyman, 1948, 1964, 1968). Some of these are often much more convenient than equation (3.61) in the analysis of actual experimental data. We return to such linked functions in Chapter 5.

Although such linkage relations are perfectly general for any system, they are especially important in biochemistry. There is good reason for this. Pure thermodynamics does not specify the magnitude of the interaction between the components of a system; it does not tell us how much the addition of component A will affect the potential, or activity, of component B. It does require, however, that the effect of A on B must bear a reciprocal relation to the effect of B on A. It is a fact of experience that the magnitude of such interactions between components is often very large in biochemical systems, as it is for oxygen and CO_2 in haemoglobin solutions. This is just another way of saying that biochemical systems are highly organized; the coupling between components is large, and the reciprocal relations are therefore of particular importance.

3.11 Ideal solutions

Ideal solutions are defined as solutions in which the activity of each component is proportional to its mole fraction; or, in terms of chemical potentials,

$$\mu_i - \mu_i^\circ = RT \ln x_i \tag{3.62}$$

No actual solution is ideal, but some are very nearly so; for instance, solutions of benzene and toluene, or of *n*-hexane and *n*-octane, approximate closely to ideal solutions. Such systems must be composed of molecules that are chemically very similar, and also not very different in size and shape.

If we mix x_1 moles of component 1 with x_2 moles of component 2, to form 1 mole of mixture, in an ideal two-component solution, the total change of Gibbs energy is

$$\begin{aligned}\Delta G_{\text{ideal}} &= x_1(\mu_1 - \mu_1^\circ) + x_2(\mu_2 - \mu_2^\circ) \\ &= x_1 RT \ln x_1 + x_2 RT \ln x_2.\end{aligned} \tag{3.63}$$

Moreover, if a solution is to remain ideal over a finite range of temperature and pressure the mixing of the components to form the solution must occur with no change of volume and no enthalpy change. This is almost intuitively obvious from the very concept of an ideal solution. The partial molar volume of each component, which determines the change of its potential with changing pressure, must be the same as the molar volume of the corresponding pure liquid, which can be taken as the standard state. The different kinds of molecules in an ideal solution must be so nearly alike that ΔH of mixing should be zero. More explicit proofs of these propositions may be found in any standard text on chemical thermodynamics.

The entropy of mixing, to form 1 mol of mixture in an ideal solution containing m components, is obtained very simply from equation (3.63) and the condition that ΔH of mixing is zero:

$$\Delta S_{\text{ideal}} = -\Delta G_{\text{ideal}}/T = -R \sum_{i=1}^{m} x_i \ln x_i. \tag{3.64}$$

We note that ΔS_{ideal} is always positive, since every one of the x_i values is less than 1. This of course is to be expected, since the mixing is a spontaneous process. (See the discussion of mixing in Section 2.5.)

The Belgian chemist Raoult in 1885 proposed a relation that must hold rigorously for ideal solutions, namely that the partial vapour pressure of any component is proportional to its mole fraction. Hence its activity is equal to its mole fraction, provided that we take pure component i as the standard state:

$$a_i = p_i/p_i^\circ = x_i. \tag{3.65}$$

This relation does not hold for most conditions in most actual solutions, but it does hold, as a limiting relation, when x_i is very close to unity. We now turn to consideration of non-ideal solutions, beginning with the limiting conditions in dilute solutions.

3.12 Dilute solutions: relation of Henry's law for the solute to Raoult's law for the solvent

It is a fact of experience that non-dissociating solutes, at high dilution, obey Henry's law; that is, the activity, and hence the partial pressure, of a volatile solute is a linear function of its mole fraction, x_2, when $x_2 \to 1$.†

For simplicity we consider a two-component system in which component 1, the solvent is at mole fraction $x_1 \simeq 1$; component 2, the solute, is at x_2, which is close to zero. We assume both components to be volatile; the case of solutes with essentially zero vapour pressure will be considered later. We define the activities of the components, as before, by the relation $a_i = p_i/p_i^\circ$, so that the activity of each component is unity for the pure compound at the same temperature. However, the assumption made for the ideal solution, that $a_i = x_i$, is no longer assumed to hold. Nevertheless, from the empirical evidence for the validity of Henry's law, we can write

$$\lim_{x_2 \to 0} (a_2/x_2) = \lim (p_2/p_2^\circ x_2) = k \tag{3.66}$$

for the solute. Here k is the Henry's law constant, which is unity for an ideal solution, but may have any value greater or less than unity in an actual solution. To see how this relation leads to Raoult's law for the solvent in such dilute solutions, we consider the general relation given in equation (3.59g):

$$\mathrm{d}(\ln a_1)/\mathrm{d}(\ln x_1) = \mathrm{d}(\ln a_2)/\,\mathrm{d}(\ln x_2). \tag{3.59h}$$

In this case, on the assumption that the components in the vapour phase behave as ideal gases, it can also be written as

$$\mathrm{d}(\ln p_1)/\mathrm{d}(\ln x_1) = \mathrm{d}(\ln p_2)/\mathrm{d}(\ln x_2). \tag{3.59h*}$$

†For solutes that dissociate – in particular those that dissociate into ions – the limiting relations depend on the number of dissociation products (ν) per mole of solute. (Thus $\nu = 2$ for NaCl, or $MgSO_4$, and 3 for $CaCl_2$). Then, to apply Raoult's law, as the concentration of solute approaches zero, if the solution contains n_1 moles of solvent and n_2 moles of solute, we must take the mole fraction of component 1 as $x_1 = n_1/(n_1 + \nu n_2)$. Van't Hoff recognized this very early (1886) and realized that the behaviour of electrolyte solutions in this respect provided powerful evidence for the Arrhenius theory of ionic dissociation. Thus the term 'mole fraction', when applied to such solutions, is ambiguous unless carefully defined. In fact the composition of such solutions is generally expressed in terms of volume concentrations, or of moles of solute per unit mass of solvent.

It is easy to see that the right-hand side of (3.59h*) must approach unity as x_2 approaches zero, for dp_2/dx_2 becomes equal to p_2/x_2 as both p_2 and x_2 approach zero. Then from (3.59h*) we have

$$\lim_{x_1 \to 1} (d(\ln p_1)/d(\ln x_1)) = 1 \tag{3.59i}$$

for the solvent. On integrating this equation we obtain

$$p_1 = p_1^\circ x_1 \tag{3.59j}$$

(if x_1 is near unity).

This is the limiting case for the application of Raoult's law to the solvent, when x_2 is given by Henry's law. Even though (3.59j) is a consequence of the fact that the solute obeys Henry's law, over a limited range near $x_2 = 0$, the result is quite general and does not depend at all on the value of the coefficient k in (3.66).

3.13 Two-component systems, with both components volatile: conventions for definition of standard states and activity coefficients

We now consider actual systems that may be far from ideal. This is of importance for most solutions of real biological interest when high concentrations of ionic and non-polar residues are introduced into water. As an example we choose the water and *n*-propanol system at 25 °C, studied by Butler, Thompson, and McLennan (1933). Figure 3.6 shows the partial vapour pressures of the two components over the entire range of composition. As theory predicts, Raoult's law holds for water when x_1 is close to 1, and for propanol when x_2 is close to 1. In these regions the actual measurements become tangential to the line drawn for an ideal solution. At all other compositions, however, the actual data lie above the ideal solution lines. For *n*-propanol, at x_2 near zero, the limiting line fits the Henry's law equation, with $k = 14.4$:

$$\lim(p_2/p_2^\circ) = 14.4\ x_2.$$

On the other hand, for water in propanol, at x_1 near zero,

$$\lim(p_1/p_1^\circ) = 3.7\ x_1.$$

Thus the deviations from Raoult's law are positive, i.e. the Henry's law constants are greater than unity. Such positive deviations are characteristic of aqueous solutions of substances containing alkyl groups, and the deviations become greater as the number of alkyl groups increases. Methanol in water shows rather small positive deviations; for ethanol they are more marked; for propanol, as seen in Figure 3.2, they are still larger. For *n*-butanol they are so large that, at intermediate compositions, the mixture separates into two phases – an upper phase containing chiefly butanol and a lower phase containing mostly water. The higher alcohols become increasingly insoluble in

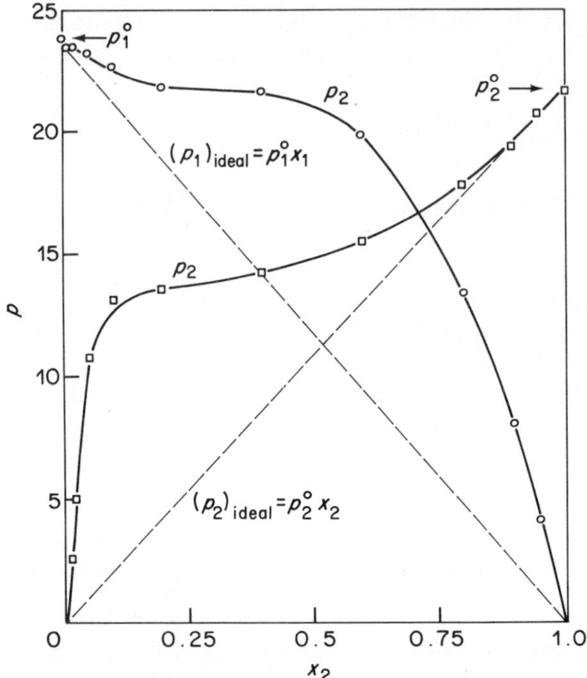

Figure 3.6 Vapour pressure diagram for the water–*n*-propanol system at 25 °C. The dashed rules show the partial vapour pressures that would be obtained if the solution were ideal. (From data of Butler *et al.*, 1933.)

water; for each added CH_2 group the Henry's law coefficient increases by a factor of the order of 3 or 4. Thus for *n*-octanol Butler *et al.* found $p_2/p_2^\circ = 12\,300\,x_2$ from solubility measurements. These effects of alkyl groups are manifestations of what are known as hydrophobic interactions (see Section 2.6).

Some solutions show *negative* deviations from Raoult's law: mixtures of acetone and chloroform furnish an example. Negative deviations indicate an attraction between the components. In this case the hydrogen of the chloroform tends to form a hydrogen bond with the negative (oxygen) end of the C—O dipole of the acetone. Positive deviations from Raoult's law indicate repulsive interactions, in the sense that a molecule in such a system tends to associate with others of its own species, rather than with those of the other component of the system.

3.13.1 Activity coefficients

In equation (3.58) we have defined the activity coefficient of component i, γ_i, as equal to a_i/x_i, and have defined pure component i (with $p_i = p_i^\circ$) as the standard state, so that $\gamma_i = p_i/p_i^\circ x_i$. Thus the activity coefficient of

n-propanol, at high dilution in aqueous solution, is 14.4 by this convention. Thus the Gibbs energy of transfer of 1 mol of propanol, from pure propanol to a very dilute aqueous solution of mole fraction x_2, is

$$\Delta G = \mu_2 - \mu_2^\circ = RT \ln a_2 = RT \ln x_2 + RT \ln \gamma_2$$
$$= \Delta G_{\text{ideal}} + RT \ln \gamma_2. \tag{3.67}$$

The term ΔG_{ideal} is always negative for such a transfer, since $x_2 = 1$ in the standard state and is always much less than 1 in the dilute solution. In this case, however, the non-ideal term, $RT \ln \gamma_2$, is large and positive, being equal to 6.62 J K^{-1} mol^{-1}.

3.13.2 An alternative standard state for the solute

Instead of choosing the pure liquid compound as the standard state with unit activity, it is often more convenient to define the standard state of the solute by the convention that $a_i = x_i$ at infinite dilution ($x_i \simeq 0$) and in the region of high dilution where Henry's law is valid. In the water–propanol system of Figure 3.6, we can find the corresponding standard state for n-propanol by drawing a tangent, at $x_2 = 0$, to the limiting curve for p_2 versus x_2, and extending the resulting straight line until it cuts the right-hand vertical axis of the figure, at $x_2 = 1$. This defines a standard state corresponding to a vapour pressure p_2^*, which for n propanol is equal to $14.4p_2^\circ$. In general $a_2 = p_2/p_2^*$ by this convention, and the activity coefficient γ_2^* is given by the equation

$$\gamma_2^* = p_2/p_2^* x_2. \tag{3.68}$$

Since γ_2^* by this convention becomes equal to unity as x_2 approaches zero, and as $p_2 = kx_2$ by Henry's law under the same conditions, it follows from (3.68) that p_2^* is equal to the Henry's law coefficient k. Thus, for pure propanol, by this convention, γ_2^*, at $x_2 = 1$, is equal to the actual vapour pressure of pure propanol (p_2°) divided by the extrapolated vapour pressure p_2^*, which is 1/14.4 = 0.0694. Thus by this convention the activity varies from 1 at infinite dilution to 0.0694 in pure propanol. By the previous convention it varies from 1 in pure propanol to 14.4 at infinite dilution in water. At any mole fraction $\gamma_2 = 14.4\ \gamma_2^*$. The values of the two activity coefficients are listed in Table 3.4 for a few values of x_2.

We note that, for a component considered as the solvent, and specifically for water in most biochemical systems, only the first of these conventions is used in practice; that is, we take the activity of water as unity in pure water at the temperature in question. For solutes at low concentration, however, the second convention is often the most convenient.

It must be understood that this convention for the standard state of the solute may not correspond to any actual state. Indeed for n-propanol in water this is clearly true. The hypothetical standard vapour pressure, p_2^*, is obviously unattainable at the experimental temperature of 25 °C. Moreover other

Table 3.4 Vapour pressure and activity coefficients of *n*-propanol in propanol–water mixtures at 25 °C, by two different conventions for the standard state. Vapour pressure of pure *n*-propanol, $p_2^\circ = 21.8$ mm Hg = 0.0287 atm = 2.83 kPa; extrapolated vapour pressure, $p_2^* = 314$ mm Hg = 0.413 atm = 40.8 kPa

x_2	p_2/p_2°	$\gamma_2 = p_2/p_2^\circ x_2$	$\gamma_2^* = p_2/p_2^* x_2$
0.0	–	14.1	(1.000)
0.01	0.123	12.3	0.854
0.02	0.232	11.6	0.805
0.05	0.495	9.9	0.69
0.20	0.624	3.12	0.217
0.40	0.652	1.63	0.113
0.60	0.712	1.19	0.083
0.80	0.816	1.02	0.071
1.00	1.000	1.00	0.0694

The data are from Butler *et al.* (1933) who took the standard state in the gas phase as a partial pressure of 1 mm Hg. However, the data tabulated here are independent of any particular choice of standard state.

Here we take the standard state in the gas phase as a partial pressure of 1 atm.

partial molar properties of the solute in this standard state – such as the partial volume, entropy, or heat capacity – are those of the solute at infinite dilution in water, not those of the pure solute. Although this definition of standard state may appear confusing at first, it is in fact an eminently practical definition, and furnishes a useful basis for calculations.

Finally we note the importance of the Gibbs–Duhem equation in its various forms – equations (3.59d)–(3.59h) inclusive – in defining the necessary relations between the changes in the activities of the two components as the composition varies. A convenient form for the equation in a two-component system is, from equation (3.59h),

$$d(\ln p_1)/d(\ln x_1) = d(\ln p_2)/d(\ln x_2).$$

This follows from equation (3.59g) since for any volatile component $d(\ln p_i) = d(\ln a_i)$. Moreover, since $x_1 + x_2 = 1$, $dx_2 = -dx_1$. From these relations it follows that

$$\frac{dp_1/dx_1}{dp_2/dx_1} = -p_1x_2/p_2x_1 \cdot \tag{3.69}$$

This relation imposes a rigorous constraint on the relative variations in p_1 and p_2 as the composition varies. If we know the value of p_1 as a function of x_1 over the whole range of composition, and if we know the value of p_2 at any one composition, equation (3.69) in principle allows us to trace out the entire curve for the vapour pressure of component 2. In this case we can check the calculated curve against the actual data. In cases that we shall shortly consider, in which the solute has effectively zero vapour pressure, we must employ the Gibbs–Duhem equation to infer the activity of the solute from that of the solvent.

3.14 Changes in Gibbs energy, enthalpy, entropy, and heat capacity for solution of hydrophobic substances in water

The hydrophobic interactions, displayed for the water–propanol system in the data of Figure 3.6, are also manifested in other thermodynamic properties of such systems – notably the enthalpy, entropy, and heat capacity and volume changes. We have already illustrated the last-mentioned in the discussion of partial molar volumes.

These properties are illustrated in Table 3.5 for the primary alcohols from methanol to *n*-pentanol. The Gibbs energy of transfer from the pure organic solute to dilute aqueous solution, given in Table 3.5 under the heading $\Delta_s G_u^\circ$, is the actual Gibbs energy minus the corresponding ΔG for an ideal solution. (This is sometimes called the unitary Gibbs energy; the subscript u is chosen for this reason.)

$$\Delta_s G_u^\circ = \Delta G^\circ - G_{\text{ideal}}^\circ = RT \ln \gamma_2 = RT \ln (p_2/p_2^\circ x_2) \quad \text{at} \quad x_2 \to 0. \tag{3.70}$$

Likewise the Gibbs energy of transfer may be taken as involving the transfer of solute from the ideal gaseous state ($p = 1$ atm) to dilute aqueous solution. This is given by

$$\Delta_h G_u^\circ = RT \ln(p_2^*/x_2) \quad \text{at} \quad x_2 \to 0. \tag{3.71}$$

Note that the standard state in the gas phase is the pure gas component at a partial pressure of 1 atm. The standard state in the liquid phase is defined by the second convention given above; that is that $a_2 = x_2$ as $x_2 \to 0$.

The enthalpies of transfer are best determined by calorimetry (see Chapter 6); it is plain from Table 3.5 that they are all negative; that is, the transfer of an organic substance, with hydrophobic groups, to water is an exothermic process at 25 °C. However, there is a large positive value of ΔC_p associated with the transfer, so that ΔH becomes steadily less negative as the temperature rises, and will presumably become positive at some transition temperature. The value of ΔH for transfer from the gas phase to dilute solution, i.e. the enthalpy of hydration, obviously involves the large negative enthalpy of condensation of the gas to the liquid state, as well as the enthalpy of solution in water. It is clear from Table 3.5 that ΔH of hydration is related to the structure of the alcohol in a much more simple and regular way than the enthalpy of solution. This is to be expected, for the enthalpy of hydration involves simply the interactions between the organic solute and water, and eliminates consideration of the interactions between the molecules in the pure organic liquid.

Table 3.5 Thermodynamic properties of some primary alcohols on transfer from the pure liquid, or from the gas phase, to dilute aqueous solution at 25 °C

Substance	$\lim \gamma_2 = p_2^{x_2 \to 0}/p_2^\circ x_2$	$\Delta_s G_u^\circ$/ kJ mol^{-1}	$\Delta_s H$/ kJ mol^{-1}	$\Delta_s S^\circ$/ J K^{-1} mol^{-1}	$\Delta_h G_u^\circ$/ kJ mol^{-1}	$\Delta_h H$/ kJ mol^{-1}	$\Delta_h S^\circ$/ J K^{-1} mol^{-1}	$C_{p_2\infty}$/ J K^{-1} mol^{-1}	ΔC_{p_2}/ J K^{-1} mol^{-1}
Methanol	1.51	1.05	− 7.24	−27.8	−3.52	−44.52	−137.5	158	97
Ethanol	3.69	3.27	−10.10	−44.6	−3.09	−52.40	−165.5	257	164
n-Propanol	14.4	6.64	−10.13	−57.2	−2.38	−57.45	−184.8	355	236
n-Butanol	52.9	9.84	− 9.28	−64.2	−1.86	−61.58	−200.2	438	300
n-Pentanol	214	–	− 7.81	–	−0.83	−64.75	−214.2	527	–

The vapour pressure data are from Butler (1937). The subscript s (in $\Delta_s G^\circ$, etc.) refers to transfer from the pure liquid alcohol, taken as standard state, to very dilute aqueous solution ($x_2 \to 0$); the subscript h denotes transfer from the gas phase, referred to a hypothetical pressure of 1 atm (101.3 kPa), to dilute aqueous solution. Thus $\Delta_s G_u^\circ = RT \ln \gamma_2$ (see equation (3.70)) and $\Delta_h G_u^\circ = RT \ln(p^*/x_2)$ (see equation (3.71 and Table 3.4). For the ΔH° and ΔS° values, see Konicek and Wadsö (1971) and Gill and Wadsö (1976). The symbol $C_{p_2\infty}$ denotes the partial molar heat capacity of the alcohol at infinite dilution in water; and $\Delta C_{p_2} = C_{p_2\infty} - C_{p_2}^\circ$, i.e. the increment in heat capacity on transferring 1 mol of pure liquid alcohol to aqueous solution at infinite dilution. These data are from Nichols, Sköld, Spink, Suurkusk, and Wadsö (1976). Note the increase in ΔC_{p_2} of 60–70 J K^{-1} mol^{-1} for each added —CH_2— group in the alcohol. For general discussions of these and related data, see Franks and Reid (1973), Franks (1975), and Edsall and McKenzie (1978).

Table 3.6 Changes in some thermodynamic properties on transfer of hydrocarbons to aqueous solution from a non-polar solvent or from the gas phase at 25 °C

	$\Delta G°$/ kJ mol^{-1}	$\Delta H°$/ kJ mol^{-1}	$\Delta S°$/ J K^{-1} mol^{-1}
Transfer from a non-polar solvent to water			
CH_4 in benzene → CH_4 in H_2O	10.91	−11.7	−75
CH_4 in CCl_4 → CH_4 in H_2O	12.1	−10.5	−75
C_2H_6 in benzene → C_2H_6 in H_2O	15.9	− 9.2	−75
C_2H_4 in benzene → C_2H_4 in H_2O	12.2	− 6.7	−63
Liquid benzene → benzene in H_2O		+ 2.1	
Liquid toluene → toluene in H_2O		+ 1.7	
Transfer from gas phase to water			
$CH_4(g)$ → CH_4 in H_2O	26.3	−13.5	−133
$C_3H_8(g)$ → C_3H_8 in H_2O	26.1	−21.3	−159
n-$C_4H_{10}(g)$ → C_4H_{10} in H_2O	26.7	−24.0	−170
Benzene (g) → benzene in H_2O	13.9	−31.6	−152
Toluene (g) → toluene in H_2O		−36.2	

Data for non-polar solvents from Kauzmann (1959); data for the gas phase from Alexander, Hill, and White (1971) and from Gill, Nichols, and Wadsö (1976).

Table 3.6 contains some similar data for hydrocarbons in water. The values for ΔH of solution for transfer of benzene or toluene molecules from the pure organic liquid to water are negative below about 23 °C, pass through zero, are already slightly positive at 25 °C, and become increasingly positive above 25 °C, due to the large positive value of ΔC_p of solution.

The standard entropy of interaction – corrected, like the Gibbs energy, for the entropy term for mixing in an ideal solution – is always negative, and becomes more negative for each added —CH_2— group in the alcohol.† Here again the increment per CH_2 group is more regular for the entropy of hydration than for the entropy of solution of the component in question. This negative entropy term, associated with hydrophobic interactions, is of profound importance for understanding the relation of water to hydrophobic groups. The interpretation of the process in molecular terms is still under debate. Here we simply set forth the thermodynamic data, and indicate their

†We emphasize again that the numerical values of $\Delta G°$ and $\Delta S°$ depend fundamentally on the choice of standard states in the two phases between which the transfer takes place. If we make a consistent choice of standard states in comparing a set of related compounds, the differences between the different substances are significant, and are largely independent of the choice of standard states. Thus the increment per CH_2 group in the various quantities in Table 3.5 is significant, and is largely independent of the particular choice of standard state; but the absolute values for any one substance are greatly dependent on the standard states chosen.

The ΔH and ΔC_p values, on the other hand, are obtainable directly from calorimetric measurements, and are essentially unaffected by the choice of standard state (see Chapter 6).

regularities, without attempting to interpret them in molecular terms. Some aspects of these phenomena are introduced in Section 2.6.

Since the properties of water and aqueous solutions are so fundamental for biochemical systems, we note the comprehensive treatise on water and aqueous systems edited by Franks (1972–82), which contains an immense amount of information by numerous contributors. See also Edsall and McKenzie (1978).

3.15 Thermodynamic properties of gases in aqueous solution

Living organisms absorb gases from, and release them to, the air, and the biochemical aspects of gas exchange are present everywhere. These include not only the intake of oxygen and release of carbon dioxide by aerobic organisms, but the evolution of methane and other hydrocarbons from certain bacteria, the action of carbon monoxide (which arises from natural sources as well as from human activities) in competing with oxygen for binding to haem proteins, and the fixation of nitrogen by legumes in cooperation with nitrogen-fixing bacteria. All these processes take place in media containing substantial quantities of water, so that it is useful to consider the solubility of gases in water and its variation with temperature. From such data we obtain information concerning such thermodynamic properties as ΔG°, ΔH°, ΔS°, and ΔC_p° for solution of these gases in water. Some solubility data are given in Table 3.7 and the thermodynamic properties for solution of the gases in water in Table 3.8. We include the noble gases helium and argon as examples of monatomic spherical gases; biologically they are interesting because they can act as anaesthetics.

Some general features of the data in Table 3.7 are immediately apparent. All these gases, except CO and N_2O, are non-polar, and even they have very small dipole moments. The solubilities are given as ratios of volume concentrations in the gas phase to volume concentration of solute in the liquid phase (Ostwald coefficients), so they are independent of the concentration units employed. The ratios are limiting values, extrapolated to zero concentration of gas, but they are essentially independent of concentration at low pressures, according to Henry's law. For all these gases except helium, the solubility coefficient decreases markedly with rising temperature. In most cases the solubility at 50 °C is less than half that at 0 °C. The decrease from 25 to 50 °C, however, is less than that from 0 to 25 °C. Most of the gases are only slightly soluble in water (and in fact are somewhat more soluble in organic solvents). Thus oxygen, for instance, at 25 °C dissolves in water at a concentration of only 0.0391 M when the concentration in the gas phase is 1 M (0.0408 atm), and the corresponding value at 40 °C is only 0.0265 M. In work on equilibrating oxygen with haemoglobin solutions, for instance, it is important to use the proper solubility coefficient for the temperature of the experiments, to calculate the concentration of free oxygen in the liquid phase (see Section 5.10).

Table 3.7 Solubilities of some gases in water between 0 and 50 °C

T/K	Solubility coefficients, $Q \times 10^3$												
	He	Ar	N_2	O_2	CO	CO_2	N_2O	CH_4	C_2H_6	C_2H_4	C_2H_2	n-C_3H_8	n-C_4H_{10}
273.15	9.40	53.6	23.8	49.0	35.5	1717	1286	57.3	99.6	–	1750	91.9	85.1
298.15	9.45	34.1	15.9	31.1	23.3	828	594	34.0	45.3	116.2	1013	36.6	29.8
313.15	9.88	28.8	14.0	26.5	20.2	605	419	28.0	33.7	92.5	816	26.1	19.9
323.15	10.32	26.7	13.4	24.8	19.1	512	–	25.7	29.4	82.5	734	22.4	16.4

Q is the Ostwald coefficient, which is the ratio of the volume concentration of substance in the gas phase to the concentration in the aqueous phase. It is therefore dimensionless, and independent of the choice of concentration units. Data from Wilhelm, Battino and Wilcock (1977).

Table 3.8 Some thermodynamic properties of gases in aqueous solution

Property	T/°C	He	Ar	N_2	O_2	CO	N_2O	CH_4	C_2H_6	C_2H_4	C_2H_2	n-C_3H_8	n-C_4H_{10}
$\Delta G°$	25	29.4	26.2	28.2	26.5	27.2	19.2	26.3	25.6	23.2	17.9	26.1	26.6
$\Delta H°$	25	−0.76	−12.27	−10.46	−12.05	−11.13	−21.4	−13.8	−19.7	−15.4	−14.8	−22.5	−26.0
$\Delta H°$	40	+0.80	−9.61	−7.11	−9.05	−8.24	−19.2	−8.2	−15.3	−12.8	−12.1	−17.0	−20.4
$\Delta S°$	25	−101	−120	−129.4	−129.2	−128	−136	−128	−152	−129	−110	−163	−176
$\Delta C_p°$	25	105	177	221	200	194	144	205	301	171	178	368	373

$\Delta G°$ and $\Delta H°$ in kilojoules per mole; $\Delta S°$ and $\Delta C_p°$ in joules per kelvin per mole. The standard state for the vapour phase is a partial pressure of 1 atm; the standard state for the liquid phase is a hypothetical state of unit mole fraction for the solute, but the actual data are based on solubilities at very low partial pressures.

Data are from Wilhelm *et al.* (1977), and are based on a critical appraisal of the best available solubility data over a range of temperatures. The data for each substance were fitted to an equation for solubility coefficient as a function of temperature. Values of $\Delta H°$ were derived by differentiation of this equation, and a second differentiation was required to obtain $\Delta C_p°$. Thus the values for the latter are less reliable than those obtained by modern calorimetric measurements; but their general trend is quite similar to the values obtained by calorimetry.

Carbon dioxide and nitrous oxide are far more soluble than the other simple gases, and the unsaturated hydrocarbons, especially acetylene, are much more soluble than the saturated hydrocarbons.

From these solubility data one can calculate values of changes in thermodynamic properties that accompany transfer of these substances from the gas phase to aqueous solution. The standard state in the gas phase is taken as a partial pressure of 1 atm, and in the solution is taken as a hypothetical mole fraction of unity for the solute. Then $\Delta G^\circ = -RT \ln (x_2/p)$, where p is in atm and x_2/p is the limiting ratio as $p \to 0$. The data are given in Table 3.8, the values of ΔH°, ΔS°, and ΔC_p° being calculated from the change of ΔG° with temperature. We give the values of ΔH° at both 25 and 40 °C, to show the marked change in enthalpy of solution with temperature. The ΔH values (except for helium) are all negative at both temperatures, but they become rapidly less negative as the temperature rises, corresponding to the very large positive values for ΔC_p° of solution.

All these gases behave like other hydrophobic substances in solution, with values of ΔG° and ΔH° that vary markedly with temperature; with generally low solubilities in water and with standard entropies of solution that are markedly negative, compared to ideal solutions; with liberation of heat when the gas dissolves in water; and with large positive partial molar heat capacities. The partial molar volumes of gases also are generally lower in water than in organic solvents. The value of V_2 for argon in benzene or CCl_4 is 43–44 cm^3 mol^{-1}; in water it decreases to 32 cm^3 mol^{-1}. For methane, on transfer from CCl_4 to water, the decrease in V_2 is -22.7 cm^3 mol^{-1} (Masterton, 1954; see also Ben-Naim, 1974, p. 320).

One gas with very special properties is carbon dioxide, a linear symmetrical molecule with two C=O bonds (C=O distance 0.116 nm) and zero dipole moment. In aqueous solution dissolved CO_2 is in equilibrium with its hydrate, carbonic acid, H_2CO_3, in which the three nearly equivalent C—O bonds form a nearly flat triangle. Since the concentration of water in the system is scarcely affected by the hydration reaction, the equilibrium constant K_h is simply

$$K_h = [H_2CO_3]/[CO_2].$$

The amount of H_2CO_3 is always very small, K_h being about 0.0025 at 25 °C. Carbonic acid is comparable in strength to formic acid (pK 3.8 at 25 °C). On the other hand the dissociation constant, as usually studied, is given (at 25 °C) by the equation

$$K_1 \text{ (carbonic acid)} = \frac{[H^+][HCO_3^-]}{[CO_2] + [H_2CO_3]} = 10^{-6.35} .$$

The equilibria involving CO_2 in the atmosphere and in the oceans and lakes, as well as the CO_2 exchanges between living organisms and their environment, are fundamental for the whole geosphere and biosphere. For further details see, for instance, Henderson (1913), Edsall and Wyman (1958, Chapter 10), and Edsall (1969). This problem is discussed further in relation to the linkage of the spontaneous reactions of CO_2 and enzyme-catalysed equilibria in Section 4.5.

3.16 Chemical potential and activity of a non-volatile solute, determined from the vapour pressure of the solvent

In solutions such as those of salts, sugars, and amino acids in water, the solute has essentially zero vapour pressure at room temperature or even at considerably higher temperature. The vapour pressure of the solvent, however, is measurable, and from it we can infer the chemical potential and activity of the solute by the Gibbs–Duhem equation in its various forms: equations (3.59a)–(3.59h). We deal here only with aqueous solutions. Most workers in this field have reported the composition of their solutions on the molality scale (in moles of solute per kilogram of H_2O) rather than in mole fractions. We will express the relations in terms of molality, and write the activity of the solute as $a_2 = \gamma_2 m$, where m is the molality of the solute. Then the Gibbs–Duhem equation becomes

$$\begin{aligned} m\,\mathrm{d}(\ln a_2) &= m\,\mathrm{d}(\ln m) + m\,\mathrm{d}(\ln \gamma_2) \\ &= -55.51\,\mathrm{d}(\ln a_1) = -55.51\,\mathrm{d}(\ln(p_1/p_1^\circ)) \end{aligned} \tag{3.72}$$

Here 55.51 is the number of moles of water in 1 kg. To integrate equation (3.72) it is convenient to introduce a quantity ϕ, commonly called the osmotic coefficient of the solute. It is defined by the equation

$$\phi = -55.51\,(\ln a_1)/m. \tag{3.73}$$

To determine ϕ we need know only the vapour pressure of the solvent and the molality of the solute. Since we know that Raoult's law holds for the solvent when its activity is near unity ($x_1 \to 1$), ϕ must approach unity as m approaches zero.† On differentiating (3.73), and making use of (3.72), we obtain

$$\mathrm{d}(\phi m) = \phi\,\mathrm{d}m + m\,\mathrm{d}\phi = m\,\mathrm{d}(\ln m) + m\,\mathrm{d}(\ln \gamma_2).$$

On dividing by m, and rearranging, this becomes

$$(\phi - 1)\,\mathrm{d}(\ln m) + \mathrm{d}\phi = \mathrm{d}(\ln \gamma_2). \tag{3.74}$$

Since $\phi \to 1$ as $m \to 0$, this becomes, on integration

$$\ln \gamma_2 = \int_0^m (\phi - 1)\,\mathrm{d}(\ln m) + (\phi - 1) = \int (\phi - 1)\,m^{-1}\mathrm{d}m + (\phi - 1). \tag{3.75}$$

†When x_1 is near unity, and $x_2 = 1 - x_1$ is very small, we can use the series expansion

$$\ln(1 - x_2) = -x_2 - x_2^2/2 - x_2^3/3 - \dots .$$

When x_2 is very small, it becomes equal to $m/55.51$ in the limit. It is then readily shown from (3.73) that ϕ must approach unity, since a_1 approaches $(1 - x_2)$ in this limit.

The integration may be carried out either graphically or analytically. The graphical integration involves plotting $(\phi - 1)/m$ against m. Both the numerator and the denominator of this expression approach zero as m goes to zero, so that high accuracy is required in the dilute solutions if the integral is to be evaluated reliably.

In these dilute solutions other methods are available for determining the potential and activity of the solvent, with much higher precision than can be obtained by direct measurement of the vapour pressure. The depression of the freezing point of the solvent, due to the presence of the solute, can be measured, by taking extreme precautions, to within an uncertainty of $\pm 2 \times 10^{-5}$ K. For most of the data in the literature, the uncertainty may be at least 10 times as great as this. Even so, the freezing point depression method provides far higher accuracy in dilute solutions for determining the activity of water and the osmotic coefficient. Few biochemists have occasion to make such measurements themselves, and we shall not discuss the procedures or the calculations involved. The calculations, and references to the experimental techniques, are given by Lewis and Randall (1961, Chapter 26) by Klotz and Rosenberg (1972, pp. 374–383), and by Harned and Owen (1958, pp. 407–423). We note that the freezing point method determines the chemical potential of the water in the solution that is in equilibrium with ice at the freezing point. What we want to know is the potential of the water, and the activity coefficient of the solute, in a solution of the same composition but at some standard temperature, usually 25 °C. This involves knowledge of the enthalpy and heat capacity changes in the solution over the temperature range from the freezing point to the standard temperature.

3.16.1 Activity of the solute from isopiestic vapour pressures of the solvent

There is a use of vapour pressure measurements, however, which can provide much higher accuracy in dilute solutions than the direct measurement of the vapour pressure, i.e. the activity of the solvent. This requires the availability of some carefully studied non-volatile solute, the activity of which has been determined with high accuracy in aqueous solution over a wide range of concentration, usually by the freezing point method, or (for electrolytes) by the electromotive force method which we discuss later. We can then set up two open containers, inside a closed vessel: one containing a solution of the standard substance, the other a solution of the substance under investigation. In the gas phase, inside the closed vessel, water vapour is in contact with both solutions. If the activity of water is higher in one solution than in the other, it will evaporate from the former solution and condense into the latter, until equilibrium is attained. The activity of the water (a_1) is then the same in both solutions. The solutions are then analysed. The value of a_1 in the solution of the standard substance, when its molality is known, can be read off from a tabulation of a_1 or ϕ values for the standard. The value of a_1 is of course the same in the solution under study as in the standard solution; the value of m for this solution is directly measured, and its osmotic coefficient is then immediately calculated from equation (3.73). The process is repeated over a range of molalities for both the standard and the unknown solution, giving a set of paired values for m (standard) and m (unknown). The calculation of the activity coefficient of the solute, from equation (3.75), then proceeds as before. To hasten attainment of equilibrium it is as well to evacuate most of the air from the vessel enclosing the two solutions; and to maintain thermal equilibrium the containers may be made of silver, or some other highly conducting material, and placed on a block of copper. Sucrose and potassium chloride solutions have often been used as standards in such measurements. This technique is commonly called the isopiestic (or isotonic) vapour pressure method. Extensive data for osmotic and activity coefficients have been obtained in this way for many salts, for urea and certain sugars, and for amino acid and peptide solutions.

3.17 The phase rule

The phase rule, first stated by Willard Gibbs, has found innumerable applications in chemistry and related fields such as geochemistry. In biochemistry it has been important for such purposes as interpreting solubility studies on proteins and providing evidence concerning the purity of protein preparations. We have already considered one special case in which the phase rule applies: the conditions imposed for the existence of two or three phases of a single pure substance in equilibrium (see equations (3.27) and (3.27a) and the following discussion). In that particular case, if three phases of a single component are in equilibrium, this can occur only at a single unique value of both pressure and temperature.

Suppose that we have a system containing n components, distributed between r phases in equilibrium. In each phase we can state the composition by taking one component as a point of reference and specifying the amount of each of the other components, relative to the unit quantity of the reference component. If one component is present in large excess, we generally call it the solvent, term it component 1, and state the concentrations of the other components relative to unit mass of component 1. In any case we then have $n - 1$ composition variables for each phase; hence there are $r(n - 1)$ such variables for all r phases. There are two additional variables, pressure and temperature.

At equilibrium, since the chemical potential of each component must be the same in all r phases, we have a set of equations. For component 1, for instance,

$$\mu_1^1 = \mu_1^2 = \ldots = \mu_1^r,$$

where the superscripts denote phases 1, 2, ..., r. This gives $r - 1$ equations, and there are similar equations for each of the n components, or $n(r - 1)$ equations in all.

Consider now a quantity v, the variance, which is the total number of variables in the system, minus the number of equations relating these

variables;
$$v = r(n - 1) + 2 - n(r - 1) = n - r + 2.$$

Here v, which is often also called the number of degrees of freedom of the system, denotes the number of variables that must still be specified before the state of the system is completely determined. For the system ice + water + water vapour, for instance, we have $n = 1, r = 3$; so $v = 0$, and the system is completely determined.

Consider a solution of n_2 moles of a pure protein, at its isoelectric point, in a solution of (say) sodium chloride (n_3 moles) in water (n_1 moles). Then there are three components. If we consider only the liquid phase, we have a variance of $3 - 1 + 2 = 4$ variables. On fixing T and p, two composition variables are left; we must then specify n_2/n_1 and n_3/n_1 and the system is completely determined. If we increase n_2 to the point where protein crystals begin to form, we thereby impose a new restriction. If equilibrium is to prevail we must have, for each of the three components, the relation

$$\mu_i\,(\text{liquid}) = \mu_i\,(\text{crystal}) \qquad (i = 1, 2, \text{or } 3).$$

This fixes the activity, and therefore the concentration, of the protein in the liquid phase. The solution is saturated; if the protein is truly a single pure component, adding still more protein cannot increase its concentration in the liquid phase. Only the crystalline phase will increase in amount. If we plot the mass of protein in the liquid phase, as ordinate, against the total mass of protein in the system, the result will be a straight line with a slope of 45°, until we reach saturation. From then on, the plot of the data should give a horizontal line. This test was used by J. H. Northrop in his studies of the purity of crystalline pepsin, trypsin, chymotrypsin, and their precursors.

In general the protein is not at its isoelectric point; it usually carries a net charge resulting from the addition to the isoelectric protein of acid or alkali. This represents an additional component. Commonly this is taken care of by specifying the pH as well as the salt concentration; although pH is not a component, it serves as an index of the amount of acid or base added. Thus, if a liquid and a crystal phase are both present, at fixed pressure, temperature, salt concentration, and pH, a pure protein should give a solubility that is independent of the amount of solid phase present. Figure 3.7, taken from a study on pepsin by Herriott, Desreux, and Northrop (1940) shows the solubility curve for pure crystalline pepsin by comparison with similar data for crude pepsin. It exemplifies the use of solubility as a criterion of purity, which was of great importance in establishing the fact that the crystallized enzyme preparations were proteins and not something else, at a time when this conclusion was widely doubted by many distinguished chemists.

3.18 Chemical potentials from various types of phase equilibria

3.18.1 Equilibration of a solute between two immiscible solvents

Consider for instance the distribution of a solute between water and a liquid

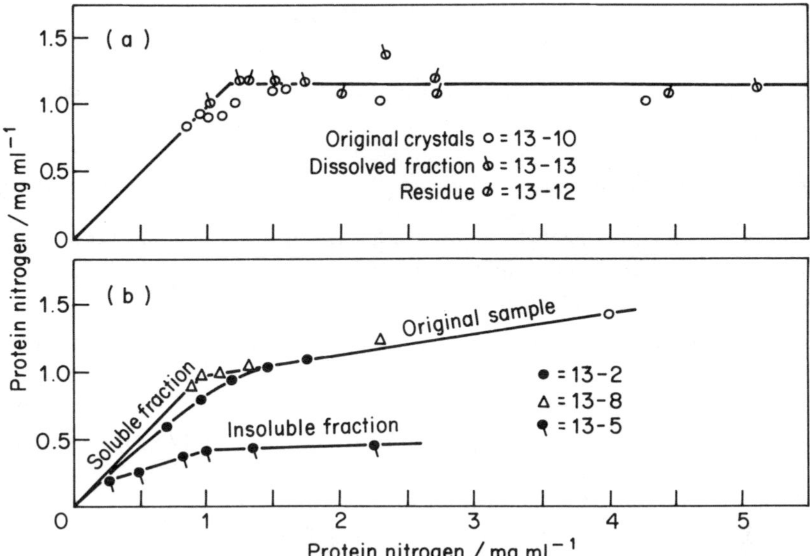

Figure 3.7 Solubility curves for (a) crystalline and (b) amorphous samples of pepsin prepared from pepsinogen. Solvent, 0.5 M saturated magnesium sulphate–0.05 M acetate buffer; pH 4.6. The theoretical curve for a system of one solid component has a slope of 1 up to the appearance of the solid phase and is parallel to the abscissa thereafter. (Redrawn from Herriott *et al.*, 1940.)

hydrocarbon, such as benzene or *n*-hexane. The chemical potential of the solute (solute component *i*) at equilibrium must be the same in both. We denote the aqueous solution by subscript w and the hydrocarbon solvent by subscript s. Then at equilibrium

$$\mu_{i,\mathrm{w}} = \mu_{i,\mathrm{s}}.$$

In the solvent

$$\mu_{i,\mathrm{s}} = \mu^{\circ}_{i,\mathrm{s}} + RT \ln a_{i,\mathrm{s}} = \mu^{\circ}_{i,\mathrm{s}} + RT \ln x_{i,\mathrm{s}} + RT \ln \gamma_{i,\mathrm{s}}. \tag{3.76}$$

Likewise in water

$$\mu_{i,\mathrm{w}} = \mu^{\circ}_{i,\mathrm{w}} + RT \ln x_{i,\mathrm{w}} + RT \ln \gamma_{i,\mathrm{w}}. \tag{3.77}$$

Then the standard Gibbs energy of transfer from solvent s to water is given by

$$\Delta G^{\circ} = \mu^{\circ}_{i,\mathrm{w}} - \mu^{\circ}_{i,\mathrm{s}} = -RT \ln (x_{i,\mathrm{w}}/x_{i,\mathrm{s}}) - RT \ln (\gamma_{i,\mathrm{w}}/\gamma_{i,\mathrm{s}}). \tag{3.78}$$

The change in standard partial molar entropy in the same process is

$$\begin{aligned} \Delta S^{\circ} &= -\partial(\mu^{\circ}_{i,\mathrm{w}} - \mu^{\circ}_{i,\mathrm{s}})/\partial T \\ &= R \ln (x_{i,\mathrm{w}}/x_{i,\mathrm{s}}) + R \ln (\gamma_{i,\mathrm{w}}/\gamma_{i,\mathrm{s}}) + RT\, \partial(\ln(x_{i,\mathrm{w}}/x_{i,\mathrm{s}}))/\partial T \\ &\quad + RT\, \partial(\ln(\gamma_{i,\mathrm{w}}/\gamma_{i,\mathrm{s}}))/\partial T. \end{aligned} \tag{3.79}$$

3.18.2 Equilibration of a component in the gas phase with a dilute aqueous solution

In this case $\mu_{i,\mathrm{w}}$ is given, as before, by equation (3.72), and $\mu_{i,\mathrm{g}}$ in the gas phase is given by

$$\mu_{i,\mathrm{g}} = \mu^{\circ}_{i,\mathrm{g}} + RT \ln p_i. \tag{3.80}$$

Here, as before in this discussion, we are assuming the gas to be ideal and we take its standard state as a partial pressure of 1 atm = 101.325 kPa.†

Then, at equilibrium between gas and aqueous solution,

$$\Delta G^{\circ} = RT \ln p_i - RT \ln x_{i,\mathrm{w}} - RT \ln \gamma_{i,\mathrm{w}} \tag{3.81}$$

and

$$\Delta S^{\circ} = R \ln (x_{i,\mathrm{w}}/p_i) + R \ln \gamma_{i,\mathrm{w}} + RT\, \partial(\ln (x_{i,\mathrm{w}}/p_i))/\partial T + RT\, \partial(\ln \gamma_{i,\mathrm{w}})/\partial T. \tag{3.82}$$

In the systems under discussion here, we define the standard state of the solute in water by the convention – see the discussion in connection with equation (3.78) – that γ_i is taken as unity at infinite dilution, in water w, and also in solvent s. (The standard states, of course, are different in the two solvents; see equations (3.76)–(3.78)). If both $x_{i,\mathrm{w}}$ and $x_{i,\mathrm{s}}$ in (3.78) are very small, it may be justifiable to assume that $\gamma_i = 1$ in both solvents. In that case (3.78) becomes

$$\mu^{\circ}_{i,\mathrm{w}} - \mu^{\circ}_{i,\mathrm{s}} = -RT \ln (x_{i,\mathrm{w}}/x_{i,\mathrm{s}}) \tag{3.83}$$

and (3.79) similarly becomes simplified by the disappearance of the activity coefficient terms. Likewise if $x_{i,\mathrm{w}}$ in (3.81) and (3.82) is very small, the terms in ln γ_i and its temperature coefficient may be ignored.

Some data on non-polar solutes in water, for systems of both these types, are given in Table 3.6. The enthalpies of solution or of hydration were sometimes determined from the temperature coefficient of solubility, and sometimes directly by calorimetry. The great insolubility of these hydrocarbons in water, as manifested by the large positive ΔG° values, is primarily determined by the large negative ΔS° values. These data, like those of Tables 3.5–3.8, manifest the characteristic features of hydrophobic interactions.

3.19 Chemical potentials and solubility

In a crystalline phase, at a given temperature and pressure, the chemical potential of any component of that phase is fixed. The crystal may be composed

†We note that the second right-hand term in (3.80) should more properly be written as $RT \ln (p/p^{\circ})$ to indicate explicitly that p represents the *ratio* of the actual partial pressure to the standard pressure of 1 atm; so we are taking the logarithm of a pure number. Ordinarily, however, we omit an explicit indication of this fact.

of a single component, as with sodium chloride or glycine, or it may consist of more than one component, as for instance $CuSO_4.5H_2O$. In any saturated solution in equilibrium with the crystalline phase, the potential of any component that is present in the crystal must be the same in the solution as in the crystal. Often we can vary the composition of the liquid – in an aqueous solution, for instance, by adding some salt to the water – without changing the composition or structure of the crystal phase at all. Then the potential of the solute in question must remain the same in all such saturated solutions. The activity will also remain the same, if we refer all measurements to the same standard state of the solute. The principles involved here are, of course, the same as in the case of the distribution of solute between two immiscible liquids, except that the potential of the solute is fixed at a single value. Thus a_i, for component i, is the same in all saturated solutions. Its solubility in the saturated solution, however, may vary greatly from one medium to another. Thus, for the potential of the solute, μ_i, when comparing solubilities in two different media, denoted by the subscripts 1 and 2, we can write

$$\mu_i - \mu_i^\circ = RT \ln a_i = RT \ln(\gamma_i c_i)_{1,\text{sat}} = RT \ln(\gamma_i c_i)_{2,\text{sat}}. \qquad (3.85)$$

The subscript 'sat' indicates that we are dealing in both cases with saturated solutions. The standard chemical potential μ_i° is defined by the convention described in the following example.

Consider a saturated solution of L-asparagine at 25 °C, which has a concentration of 0.186 M in water, whereas a saturated solution in ethanol is 2.3×10^{-5} M. L-Asparagine is thus nearly 8100 times as soluble in water as in ethanol; yet its chemical potential in the two saturated solutions is the same, since both are in equilibrium with the same crystalline phase. If we denote L-asparagine as component 2, using superscript ^{A} to denote the alcoholic solution and superscript ^{w} for the aqueous solution, at saturation we must have

$$(\mu_2^{\text{A}})_{\text{sat}} = (\mu_2^{\text{w}})_{\text{sat}} = RT \ln (a_2^{\text{w}})_{\text{sat}} = RT \ln (a_2^{\text{A}})_{\text{sat}}, \qquad (3.86a)$$

$$(a_2^{\text{w}})_{\text{sat}} = (c_2^{\text{w}})_{\text{sat}} (\gamma_2^{\text{w}}) = (a_2^{\text{A}})_{\text{sat}} = (c_2^{\text{A}})_{\text{sat}} (\gamma_2^{\text{A}})_{\text{sat}}. \qquad (3.86b)$$

Here the c_i are molar concentrations and the γ_i are activity coefficients. The fact that we have set a_2, as well as μ_2, equal in the two saturated solutions means that we have chosen the same standard states for both. (Note the difference between this assumption and that made in the preceding section, where we chose to set up different standard states for the solute in two different media.) It is convenient in this case to define the standard state by setting $a_2^{\text{w}} = c_2^{\text{w}}$ in a very dilute aqueous solution; thus γ_2^{w} in such a solution is unity by definition. The activity coefficient of an amino acid such as L-asparagine in water changes little between 0 and 0.186 M; thus we may, with little error, take $(\gamma_2^{\text{w}})_{\text{sat}} = 1$ in such a solution. Likewise, in the extremely dilute solution of asparagine in ethanol, we can take $(\gamma_2^{\text{A}})_{\text{sat}} = (\gamma_2^{\text{A}})_{m=0}$; then (3.86b) becomes

$$(\gamma_2^{\text{A}})_{\text{sat}} = (c_2^{\text{w}})_{\text{sat}}/(c_2^{\text{A}})_{\text{sat}} = 8100.$$

Consider two dilute solutions of asparagine at the same molar concentration, in water and in ethanol: $c_2^{\mathrm{w}} = c_2^{\mathrm{A}}$. Then, if we transfer an infinitesimal amount of asparagine at 25 °C from water to ethanol, the increment in Gibbs energy, *per mole of asparagine transferred*, is

$$\Delta G = RT \ln (\gamma_2^{\mathrm{A}}) = RT \ln 8100 = 22.31 \text{ kJ mol}^{-1} = 5.33 \text{ kcal mol}^{-1}.$$

This large positive Gibbs energy of transfer is due chiefly to electrostatic effects. The dielectric constant of water (78.5 at 25 °C) is more than three times as high as that of ethanol. The structure of L-asparagine is

$$H_2N(O{=})C-CH_2-CH(NH_3^+)COO^-$$

The charged $-NH_3^+$ and $-COO^-$ groups form a dipole of high electric moment, the two centres of charge being separated by about 3×10^{-8} cm, thus giving a dipole moment of $3 \times 10^{-8} \times 4.8 \times 10^{-10}$ or close to 15×10^{-18} e.s.u. cm = 15 D. Also the $-CONH_2$ group, though it carries no net charge, is strongly polar, with a moment of the order of 3.5 D. Such highly polar structures acquire a high Gibbs energy in media of low dielectric constant, whereas they are stabilized in the presence of water molecules, which are strongly polar, and readily form hydrogen bonds to other polar charged groups. These molecular properties of water, and its small size, allow many dipoles to pack into a given space, and form hydrogen bonds with one another; as a result liquid water possesses such a high dielectric constant.

A comparison of the relative solubilities of L-asparagine and L-leucine is revealing. The solubility of L-leucine in water (0.171 M) is slightly less than that of asparagine; but its solubility in ethanol is 1.28×10^{-3} M, or 69 times as great as that of asparagine. This is understandable when we consider the structure of leucine:

$$(CH_3)_2CH-CH_2-CH(NH_3^+)COO^-$$

Leucine, like all the α-amino acids, contains the dipole ($\mu = 15$ D) involving the charged amino and carboxyl groups, but the four-carbon side chain is entirely non-polar and therefore hydrophobic. The dipole is the predominating factor, so that leucine is in fact more soluble in water than in ethanol; but the purely non-polar side chain, in contrast to the polar amide group in asparagine, causes leucine to be a little less soluble than asparagine in water and much more soluble in ethanol.

We emphasize that, for thermodynamic studies, it is primarily solubility *ratios* that interest us – not the absolute value of the solubility of a substance in any one medium. The absolute solubility at a given temperature depends, not only on the interaction between the solvent and the solute, but also on the strength of the forces between the molecules (or ions) in the crystal lattice of the solid phase. The stronger the intermolecular (or interionic) forces in the crystal, the lower in general will be the solubility. However, if the same solid phase, in different experiments, is equilibrated with two different solvents, at constant temperature and pressure, then such a pair of solubility measurements is equivalent to measuring the distribution coefficient of the solute between the two solvents. When we determine such a solubility ratio, the crystal lattice energy contribution cancels out, and we can determine the change in chemical potential when the solute is transferred from one medium to the other.

Such solubility ratios, for a series of related compounds, can often be correlated with structure. As a simple example, Table 3.9 lists the solubilities of several amino acids with differing length of side chain. The ratio of solubility in water to solubility in ethanol ranges from 7400 for glycine to 83 for α-aminocaproic acid. There is a steady progression; this ratio diminishes by a factor of the order of 3 for each additional —CH_2— group in the side chain, or by 3.5 kJ mol^{-1} per —CH_2— group in chemical potential (Tanford, 1980). This effect of hydrocarbon side chains holds approximately for many other systems, involving a transfer of substances containing hydrophobic groups between water and organic solvents. Similar systematic relations can be formulated for the influence of polar groups in the solute; these effects, of course, are of opposite sign to those of non-polar groups.

3.20 Solutions of ions and their interactions with other compounds

Ionic solutes exert long range electrostatic forces on one another and on other compounds present in solutions. The resulting effects on the chemical potentials of the substances involved are of major importance. Thus the addition of salts to aqueous solutions of simple gases, or of uncharged organic compounds, generally causes the solubility of such compounds to decrease. Empirically the change of solubility of an uncharged solute, as a function of the salt concentration, is often found to be described by the simple relation

$$\log(S/S_0) = -\log \gamma = -K_s C_s. \qquad (3.87)$$

Here S_0 is the solubility of the solute in the absence of salt, S is the solubility in the presence of salt at molar concentration C_s, and K_s is commonly called a salting-out constant. Equation (3.87) often holds reasonably well up to C_s values of 2 M or even more. A fairly typical example is given by the solubility of acetone in the presence of various salts, which is shown in Figure 3.8; the data are taken from the work of Philip Gross and his collaborators. They determined the amount of acetone dissolved when the partial pressure of acetone was held at a fixed level in the vapour phase above the solution, so that the chemical potential of acetone in the solution was the same in all experiments. Thus the activity coefficient (γ) of acetone in the solution is the reciprocal of the solubility ratio (S/S_0) in equation (3.87).

Table 3.9 Solubilities of a homologous series of α-amino acids $^{+}H_3NCHRCOO^{-}$ at 25 °C

R	Amino acid	$S[H_2O]$/ mol dm^{-3}	$S[C_2H_5OH]$/ mol dm^{-3}	$S[C_2H_5OH]/S[H_2O] = Q$	log Q	$\Delta \log Q/\Delta n_{CH_2}$
H	Glycine	2.89	3.9×10^{-4}	7400	3.87	
CH_3	α-Alanine	1.66	7.6×10^{-4}	2180	3.37	0.50
C_2H_5	α-Amino-*n*-butyric acid	1.80	26×10^{-4}	690	2.84	0.53
C_4H_9	α-Amino-*n*-caproic acid	0.0866	10.4×10^{-4}	83	1.92	0.46

Data from Cohn and Edsall (1943, Chapter 9). *S* denotes solubility in the given solvent.

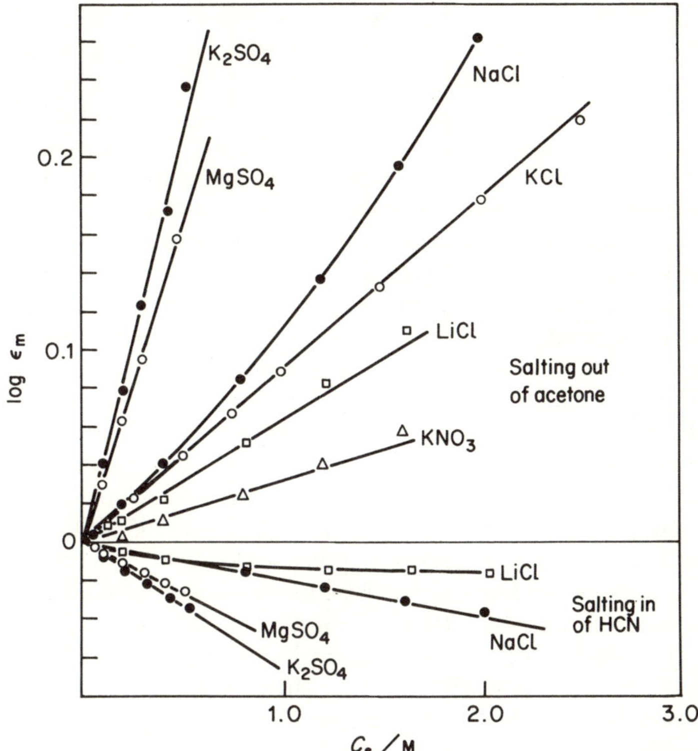

Figure 3.8 The salting out of acetone, which lowers the dielectric constant of water, and the salting in of HCN, which raises the dielectric constant. (From Gross and Schwartz, 1930, and Gross and Iser, 1930.)

The data of Figure 3.8 show that all the added salts decrease the solubility of acetone, i.e. they increase the activity coefficient. The process is denoted as 'salting out'. The values of K_s in (3.87), however, differ greatly for the different salts. K_s is roughly 0.03 for KNO_3, whereas it is close to 0.40 for K_2SO_4. These data are on the whole typical for the relative effectiveness of different salts in salting out various inorganic and organic compounds from water – including simple gases, such as hydrogen, oxygen, nitrogen, neon, and argon. The most effective salting out agents are sulphates and phosphates of sodium and potassium: chlorides are less effective; nitrates still less so. Iodides and thiocyanates have very little salting out action; indeed they sometimes increase the solubility of uncharged solutes to some extent – a process commonly called 'salting in'.

Gases are generally salted out of water. For oxygen, for instance, K_s is 0.14 for NaCl and 0.13 for KCl. For CO_2, K_s is 0.101 for NaCl and 0.073 for KCl at 25 °C (Long and McDevit, 1952). For oxygen this means that in 0.15 M NaCl, with an ionic strength similar to that of the red cell, the solubility is about 5% lower than it is in pure water. This should be taken account of in accurate work on such problems as oxygen–haemoglobin interactions, but it is a less important factor than the variation of the solubility coefficient with temperature, which we have already discussed.

The lower portion of Figure 3.8 shows the effect of several salts on the solubility of hydrocyanic acid in water. Here we see that the same salts that decrease the solubility of acetone *increase* the solubility of HCN. Acetone, when dissolved in water, decreases the dielectric constant; HCN increases it. Debye (1927) interpreted these phenomena in terms of the intense electric field around the ions, which falls off rapidly with distance from the ions. In such an inhomogeneous field, strongly polar molecules, of small volume, like water, tend to cluster around the ions, where the field strength is highest. Molecules containing non-polar groups, like acetone, are 'squeezed out' from the neighbourhood of the ions, and their solubility therefore decreases. HCN, on the other hand, is a small molecule of even higher electric moment than water; it is thus attracted around the ions, and is salted in. This simple picture is not altogether adequate to explain all the complexities of the data, but it certainly contains a large measure of truth. Long and McDevit (1952) have provided a thoughtful, critical review of the work in this field.

Slightly soluble salts commonly become more soluble in the presence of low concentrations of other salts without a common ion. Thallous chloride, barium iodate, and a variety of complex salts of cobalt compounds, studied by Brönsted (Brönsted 1920, 1922, 1923; Brönsted and LaMer, 1924) are examples of such salts (see also Scatchard, 1976, pp. 93–95). This increase in solubility reflects a decrease in the activity coefficient of the slightly soluble salt; its activity is held constant, as the total salt concentration changes, by equilibration with the constant solid phase. To keep the activity constant, the concentration of dissolved salt at saturation must go up as the activity coefficient goes down.

These effects represent one aspect of the electrostatic interactions between the ions, as formulated by Debye and Hückel (1923). A charged ion tends to attract ions of opposite charge around it, and to repel ions of like charge. Thus each ion is surrounded by a highly mobile 'ion atmosphere', of predominantly opposite sign of charge. This has the effect of partially neutralizing the charge on the central ion; this reduces its electrostatic Gibbs energy, and correspondingly reduces its activity coefficient. At low salt concentrations the mean activity coefficient of the ions of the slightly soluble salt ($\gamma_\pm$) is given by the equation

$$\log_{10} \gamma_\pm = Z_+Z_- \, PI^{1/2}/(1 + QaI^{1/2}). \tag{3.88}$$

Here Z_+ and Z_- are the valences (charges) of the cation and anion of the dissolved salt, and I is the ionic strength, defined by the relation

$$I = \tfrac{1}{2} \sum c_i Z_i^2, \tag{3.89}$$

where the summation is taken over all ions in the solution. The factors P and Q are functions of the electronic charge unit, and of the product (DT) of the dielectric constant of the solvent and the absolute temperature. P is proportional to $(DT)^{-3/2}$ and Q is proportional to $(DT)^{-1/2}$. Thus both these terms increase as D decreases at constant temperature, reflecting the fact that electrostatic interactions increase as D decreases. If the solvent is water, however, the product DT is fairly independent of temperature over a range around 25 °C, for D decreases as T increases. The factor a in (3.88) is proportional to the 'collision diameter' of the ions; for a pair of small ions it is of the order of 0.3 nm; for a small ion interacting with a spherical protein of radius R ångstroms, it is of the order of $R + 2$.

For aqueous solutions, around 25 °C, (3.88) becomes equation

$$\log_{10} \gamma_{\pm} = Z_+Z_- \,(0.50)\, I^{1/2}/(1 + 0.33\, aI^{1/2}), \qquad (3.88a)$$

where a is in ångstroms. This equation holds well only at low I values, of the order of 0.1 or less. Note that Z_+Z_- is always negative (it is -1 for a uni-univalent salt) so that $\log \gamma_{\pm}$ is always negative under conditions where (3.88a) holds. At higher I values other effects, such as salting out, become important. They can often be approximated by introducing a salting out term, proportional to the first power of I $(+K_sI)$, and opposite in sign to the term on the right of (3.88a).

The significance of the ionic strength, for interactions between ions, was first recognized empirically by G. N. Lewis in 1921; the work of Debye and Hückel (1923) established it on a theoretical foundation.†

Interactions between ions and dipolar ions, such as the amino acids and peptides, are also determined in large measure by electrostatic forces. Increase of ionic strength increases the solubility of dipolar ions such as glycine, asparagine, and cystine at low ionic strength. For example, Figure 3.9 shows the solubility of cystine in solutions of several salts. At low ionic strengths, all the salts increase the solubility of the amino acid, but the slopes of the 'salting in' curves are very different. The most effective salting in is by calcium chloride; the least by sodium sulphate. At higher ionic strengths the curves for ammonium and sodium sulphates pass through a maximum, and at still higher ionic strengths cystine is salted out by these salts. At $I = 12$, in ammonium sulphate, cystine is less soluble than it is in the absence of salt. The slope of the salting out curve, at these high ionic strengths, is similar to those for acetone in Figure 3.8.

Not all dipolar ions become more soluble in the presence of added salts, even at low ionic strengths. Leucine, with its non-polar side chain, for instance, is salted out, even at low ionic strengths, by sodium and potassium chlorides; although addition of calcium chloride to leucine solutions increases the

†For a discussion of experimental data, and their relation to the Debye–Hückel theory, see for instance Lewis and Randall, (1961, Chapters 22 and 23). Edsall and Wyman (1958, Chapter 5), give a fairly simple discussion of the basis of the theory. More advanced and rigorous discussions are given by Harned and Owen (1958, Chapters 1, 2, and 3), and by Scatchard (1976). Note that some of the earlier treatments define the ionic strength as $I/2$ where we have denoted it by I. This factor of 2 is important in comparing the work of different authors.

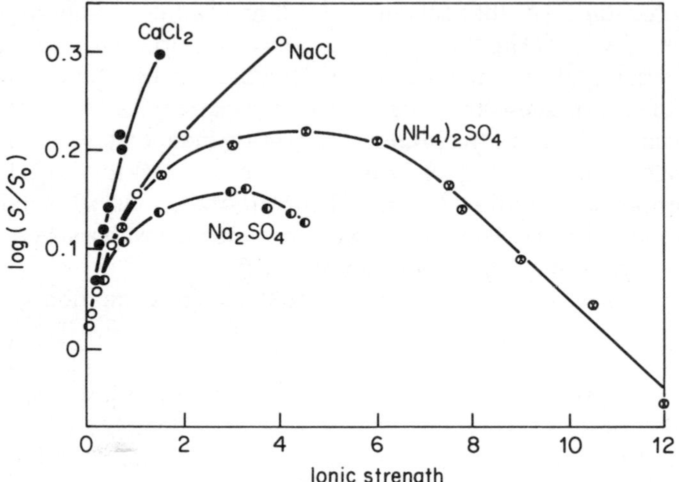

Figure 3.9 Solubility of cysteine in aqueous salt solutions. (Redrawn from Edsall and Wyman, 1958.)

solubility somewhat. In any case, the limiting slope of the solubility curve for a dipolar ion in salt solution is proportional to the first power of the ionic strength (compare equation (3.87)) not to its square root, as in equation (3.88). The square root term appears only when both the interacting substances are ions carrying a net charge (p. 94). Proteins are also salted in and out by variation of the salt concentration.† Figure 3.10, taken from the work of Green (1931), shows the solubility of horse carboxyhaemoglobin as a function of ionic strength, in several different salts. Qualitatively the curves look rather like those for cystine, but the magnitude of the effects of salts is far greater for haemoglobin (note the abscissa and ordinate scales in the two figures). The value of K_s for cystine in ammonium sulphate is 0.05; for haemoglobin in the same salt it is 0.71, 14 times as great. In a logarithmic expression this makes an enormous difference. In salting out of cystine, an increase of 6 M in ionic strength decreases solubility to approximately half its original value; for haemoglobin an increase of 5 M in ionic strength decreases the solubility by a factor of approximately 3000!

The salting in and salting out effects are thus far greater for macromolecules, such as proteins, than for small molecules. Since the salting out coefficients of different proteins, as well as their actual solubilities in dilute salt solutions, are very different, the fractionation of proteins by differential salting out was employed as early as about 1850, long before similar phenomena were recognized for the amino acids. The salting in effect, which is apparent from Figure 3.10 for haemoglobin at low ionic strength, is much more pronounced for some other globulins; for instance β-lactoglobulin from cow's milk is about

†For further details concerning the interactions of ions and dipolar ions, see Cohn and Edsall (1943, Chapters 11, 12 and 24) and Edsall and Wyman (1958, Chapter 5).

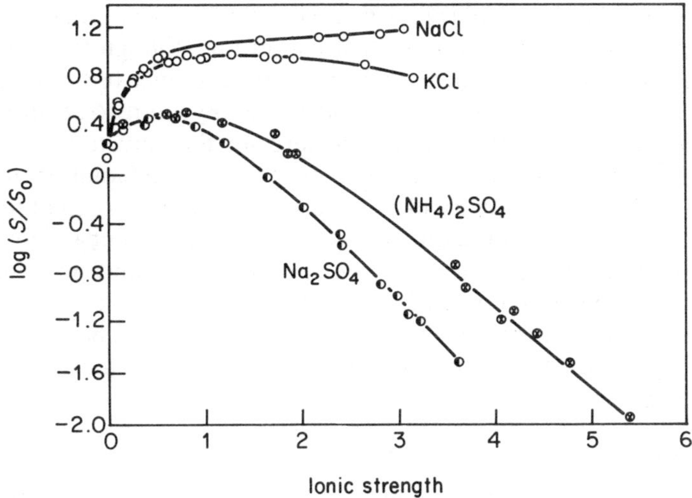

Figure 3.10 The solubility of horse carboxyhaemoglobin in salt solutions of varying ionic strengths. (From data of Green, 1931; redrawn from Cohn and Edsall, 1943.)

100 times as soluble in 0.1 M NaCl as in pure water.

In spite of the great interest and importance of these phenomena for proteins, we must be careful not to draw thermodynamic inferences unjustifiably from solubility data on protein solutions in equilibrium with crystals. Protein crystals contain large amounts of water, some of it rather strongly bound, some relatively free, in the interstices between the protein molecules in the crystal. If the liquid phase contains salt ions, or small uncharged molecules, these are generally present in the crystal phase also, although the ratio of salt to water is generally different in the crystal and in the liquid. Thus, as we vary the ionic strength, we also vary the ratio of salt to water in the crystal. Hence, since the composition of the crystal is not constant in the various experiments, we are not justified in assuming that the chemical potential of the protein in the crystal is constant, as we generally can in studies on amino acid or other simpler crystals.

3.21 Osmotic pressure: chemical potentials and molecular weight determinations

Osmotic pressure arises when two liquid phases are separated by a semipermeable membrane; that is a membrane that is permeable to some components of the solution but not to others. We are concerned here with membranes that hold back certain solutes selectively, primarily on the basis of size. It is relatively easy to obtain membranes that will allow water, simple ions of salts, and molecules of low molecular mass to pass through freely, while holding back proteins of molecular mass above (say) 25 000 daltons. Carefully prepared membranes can be made that will hold back macromolecules of the order of 10 000 daltons, while allowing reversible passage to small molecules.

It is primarily aqueous solutions that concern us, though the fundamental equations apply to solutions in any solvent, in the presence of a semipermeable membrane. We have already seen that addition of any solute to a solution lowers the activity, and hence the vapour pressure, of the solvent. In very dilute solution, Raoult's law holds, and the activity of the solvent (component 1) in the solution – if we define activity of pure solvent as unity – is equal to its mole fraction (x_1). We consider the simplest case, a two-component aqueous system, with a membrane permeable to water (component 1) and impermeable to component 2. The solution inside the membrane contains component 2; the liquid outside the membrane is pure water. Then at atmospheric pressure the activity of water in the inner solution, which we take as the relative partial pressure, is given by

$$a_1 = p_1/p_1^\circ = x_1. \tag{3.90}$$

In the pure water outside, $p_1 = p_1^\circ$ and the activity is unity. If the total pressure is equal on the two sides, equilibrium cannot exist; water will flow from the outer to the inner solution, and will continue to flow, since a_1 will always be less than unity, though it approaches unity as the solution becomes more and more dilute.

We can, however, establish an equilibrium by applying increased pressure to the inner solution (or by decreasing the pressure on the outer liquid). Under increased hydrostatic pressure the chemical potential of the water in the inner solution increases, by the fundamental relation

$$\left(\frac{\partial \mu_1}{\partial p}\right)_T = RT\left(\frac{\partial(\ln a_1)}{\partial p}\right)_T = V_1, \tag{3.91}$$

where V_1 is the partial molar volume of water. Thus, by increasing the pressure, we increase μ_1 (and a_1) since V_1 is positive, and in a dilute aqueous solution it is very nearly equal to the molar volume of pure water. By applying sufficient pressure we can increase a_1, inside the membrane, from x_1 to unity, when the system will be in balance, with no tendency for a net flow of water across the membrane. Thus we integrate equation (3.91) from $a_1 = x_1$ to $a_1 = 1$:

$$RT\int_{x_1}^{1} \mathrm{d}(\ln a_1) = \int_{p}^{p+\Pi} V_1\,\mathrm{d}p = V_1\,\Pi. \tag{3.92}$$

Here p is atmospheric pressure and Π is by definition the osmotic pressure of the solution, i.e. the pressure required to equalize the activity of water on the two sides. The definite integral on the left of (3.92) then has the value

$$\Pi V_1 = -RT \ln a_1 \simeq -RT \ln x_1. \tag{3.93}$$

This is the fundamental equation of osmotic pressure. Since we are dealing with a two-component system, $x_1 + x_2 = 1$ and

$$\Pi V_1 = -RT \ln(1 - x_2) = RT(x_2 + x_2^2/2 + x_2^3/3 + \dots), \tag{3.94}$$

where we have made use of the series expansion of $\ln(1 - x_2)$. In dilute aqueous solution the first term of the series, x_2, is much larger than any of the others. Even at the relatively high concentration of 1 mol solute per kilogram of H_2O ($x_2 = 0.018$) the second term in the series is less than 1% of the first; and we are considering solutions much more dilute than this. We are therefore justified in dropping all but the first term in the series expansion, and (3.94) becomes

$$\Pi = \frac{RT\,x_2}{V_1} = \frac{RT}{V_1}\,\frac{n_2}{(n_1 + n_2)} \simeq \frac{RT\,n_2}{n_1 V_1}\,. \tag{3.95}$$

If we choose the number of moles (n_1) of solvent so that $n_1 V_1$ is equal to 1 litre, then n_2 is the number of moles of component 2 per litre of solvent, which at high dilution becomes essentially identical with the number of moles of component 2 per litre of solution, i.e. with the molar concentration c_2. Thus (3.95) becomes

$$\lim_{c_2 \to 0} \left(\frac{\Pi}{c_2}\right) = RT. \tag{3.96}$$

This is the limiting law of van't Hoff. It shows that, at very high dilution, the osmotic pressure of a non-diffusible solute (i.e. a solute to which the membrane is impermeable) is numerically the same as the pressure of a perfect gas at the same molar concentration (c_2) of gas molecules. The thermodynamic basis for the phenomenon of osmotic pressure is of course quite different from that determining the pressure produced by the bombardment of gas molecules on the walls of a container; but the numerical equivalence of the two expressions provides a convenient basis for calculating the order of magnitude of the osmotic pressure to be expected, for a substance of known molecular weight.

Osmotic pressure measurements have been widely used for determining molecular weights of substances that cannot pass through suitably devised membranes. We can of course determine the weight concentration of solute, g_2, in kilograms per cubic metre, by direct analysis, and $g_2 = c_2 M_2$, where M_2 is the molecular weight. Thus (3.96) becomes

$$\lim_{g_2 \to 0} \left(\frac{\Pi}{RTg_2}\right) = \frac{1}{M_2}\,. \tag{3.97}$$

We must consider the order of magnitude of the Π value to be expected for molecules of a given molecular weight. One mole of a perfect gas, at 0 °C, confined in a volume of 1 litre, should give a pressure of 22.4 atm. This is the 'ideal osmotic pressure' of a solute at this concentration and temperature. Suppose we are studying a protein of molecular mass 10^5 daltons at a concentration of 10 g l^{-1} and at 0 °C. This corresponds to $c_2 = 10^{-4}$ M, which should give an ideal osmotic pressure of 2.24×10^{-3} atm, which is equivalent to 1.70 mm Hg, or 23.2 mm H_2O. If we make the pressure reading with a water manometer, it is quite possible to read the pressure with a probable error no greater than ±0.2 mm; indeed with careful technique the error could be considerably reduced. In such a solution, therefore, the probable error could be reduced below 1%. If the molecular mass were 10^6 daltons, the percentage error would be 10 times as great. For very large macromolecules, therefore, osmotic pressure is not a very accurate method of molecular weight determination.

In fact, of course, van't Hoff's law is a limiting law. Actual measurements of osmotic pressure over a range of concentrations are almost always describable by expanding the right-hand side of (3.97) into a power series in g_2:

$$\frac{\Pi}{RTg_2} = \frac{1}{M_2} + Bg_2 + Cg_2^2 + \ldots \tag{3.98}$$

We then plot Π/RTg_2 against g_2. This plot, for a dilute solution, should give a straight line of slope B and intercept M_2^{-1}.

The terms B and C in (3.98) are commonly called the second and third virial coefficients. Usually the term in B is the only one we need to consider. In general it is a function of the 'effective volume' of the solute molecules; that is, of the region around the centre of such a molecule into which solvent cannot penetrate. Attractive forces between the molecules tend to diminish B; repulsive forces to increase it. If the attractive forces are sufficiently large, B may become negative. Such a solution may be on the verge of precipitation, with separation of the solute as a separate phase.

Our discussion, up to this point, has assumed a two-component system – water plus one solute. Actual measurements practically always involve at least one more component, which may for instance be a simple salt. If the protein, or other macromolecule, carries no net electric charge, and if the third component can freely cross the membrane, this offers no real complication. The water (component 1) and the third component can together be regarded as a single solvent, and the argument leading to equations (3.95)–(3.97) is essentially unchanged.

If the macromolecule carries a net charge, however, a different situation arises. Suppose, to take a simple case, that component 2 is a protein at molar concentration c_2, and that the mean net charge of the protein molecules, in proton units is Z_2. (Note that Z_2 is a signed quantity). Suppose that we have added a single salt, say $Na^+ Cl^-$, to the system. Inside the membrane the concentrations of Na^+ and of Cl^- cannot be equal, since this liquid phase must be electrically neutral as a whole. The condition for neutrality is

$$[Na^+] + Z_2c_2 = [Cl^-]. \tag{3.99}$$

Thus, if Z_2 is positive, $[Cl^-] > [Na^+]$. If Z_2 is negative, $[Na^+] > [Cl^-]$. Outside the membrane there is no protein, and $[Na^+] = [Cl^-]$ to preserve neutrality. H^+ and OH^- ions are also present.

The salt ions move freely across the membrane but, to maintain electrical neutrality, any net transfer of the cation must be accompanied by an equivalent transfer of anion. At equilibrium we must have $\Delta G = 0$ for a transfer of an infinitesimal amount of Na^+ plus Cl^- ions across the membrane. This means that the *total potential of the sum of the two ions* must, to a close approximation, be the same on both sides of the membrane:

$$(\mu[Na^+] + \mu[Cl^-])_{\text{inside}} = (\mu[Na^+] + \mu[Cl^-])_{\text{outside}}.$$

Converting from chemical potentials to activities, and remembering that the sum of two potentials corresponds to the product of the corresponding activities, we have

$$(a_{Cl}\, a_{Na})_i = (a_{Cl}\, a_{Na})_o. \tag{3.100}$$

Generally, to a good approximation, we can replace activity ratios by concentration ratios, and on doing this, and rearranging, we obtain

$$\frac{[Na^+]_i}{[Na^+]_o} = \frac{[Cl^-]_o}{[Cl^-]_i}. \tag{3.100a}$$

This equation is, for monovalent ions, a statement of what is known as the Gibbs–Donnan equilibrium. Willard Gibbs formulated it in 1876, in very general terms, before the existence of ions in solution was known; his treatment dealt simply with the general case in which, for any reason whatever, two constituents of the system were somehow coupled, so that they were required to cross the membrane in equivalent amounts. In 1911 F. G. Donnan, unaware that Gibbs had already given the answer, derived an equation for distribution of ions, of which (3.100a) is a special case. In 1923 G. S. Adair was the first to call attention to the treatment by Gibbs.

The combination of (3.99) and (3.100a) provides the basis for calculating the additional osmotic pressure due to the unequal distribution of diffusible ions, for a solution dilute enough to allow the assumption that van't Hoff's equation holds. We calculate Π by adding together the molar concentrations of all solutes inside the membrane – both macromolecules and salt ions – doing the same for the solution outside (where only salt ions are present) and taking the difference between the two. This gives the equation

$$\Pi = RT(c_2 + [Na^+]_i + [Cl^-]_i - [Na^+]_o - [Cl^-]_o). \tag{3.101}$$

Of course, in the outer solution, $[Na^+]_o = [Cl^-]_o = c_3$. The term RTc_2 gives the Π value for the isoelectric macromolecule. The algebraic sum of the terms involving the salt ions represents the added pressure due to the Gibbs–Donnan effect.

It is easy to see that this Gibbs–Donnan term is always positive. By (3.100) we know that the $[Na^+][Cl^-]$ product is the same inside and outside, but the terms of the product are unequal on the inside and equal on the outside. Geometrically the 'inside' product corresponds to a rectangle with sides equal to the values of $[Na^+]_i$ and $[Cl^-]_i$. The 'outside' product corresponds to a square, with the same area as the rectangle. The perimeter of a square is always less than that of any rectangle of equal area, and the Gibbs–Donnan term in (3.101) corresponds to half the perimeter of the 'inside' area minus half the perimeter of the 'outside' area. Hence this term is always positive (or zero) whether the charge on the macromolecule is positive or negative.

The quantitative calculation of the Π value in (3.101) has been given by Adair (1925b, 1928), Scatchard, Batchelder and Brown (1944), Wagner (1949), Edsall (1953), and Scatchard (1976). Here we present only the final result, denoting the salt concentration in the outer solution as m_3:

$$\frac{\Pi}{RTg_2} = \frac{1}{M_2} + \frac{Z_2^2 g_2}{4M_2^2 m_3} + \dots \quad \text{or} \quad \frac{\Pi}{RT} = \frac{g_2}{M_2} + \frac{(Z_2 g_2)^2}{4M_2^2 m_3} + \dots .$$

Thus the Gibbs–Donnan term is proportional to the square of the charge on the macromolecule, and inversely proportional to the salt concentration in the outer solution. Therefore, by adding more salt to the solution, we can reduce the magnitude of this term. In any case a plot of Π/RTg_2 against g_2 should give an intercept equal to M_2^{-1} on the ordinate axis, and thus should determine the true molecular weight. Scatchard (1946) has given a detailed and rigorous proof of this important point, and has developed a general theory of osmotic pressure in multicomponent systems.

3.22 Chemical potentials and electromotive force measurements

For electrolyte solutions, electromotive force measurements provide one of the most powerful methods of determining chemical potentials. Electrical work in a galvanic cell is proportional to the amount of electricity that flows through the cell multiplied by the voltage. If the current flows at a finite rate, however, the voltage soon drops somewhat as products of the electrochemical reaction build up. To obtain the maximum possible voltage, and therefore the maximum work, it is necessary to operate under essentially reversible conditions; then the rate of current flow is virtually zero, the electromotive force (e.m.f) of the cell under study being almost exactly balanced by an external e.m.f. from a standard cell of known voltage; that is, we set up a potentiometer.

The use of such a cell in determining the chemical potential of an electrolyte is perhaps easiest to visualize in terms of a double cell, with the electrolyte – we take HCl as an example – present in both compartments of the cell, but at different concentrations. Then the result of the flow of current is simply to transfer the electrolyte from one compartment to the other.

We can set up such a cell, for two HCl solutions at concentrations c_1 and c_2, by an arrangement represented diagrammatically by

$$\text{Pt}, \text{H}_2 \mid \text{HCl}\ (c_1) \mid \text{AgCl}, \text{Ag}, \text{AgCl} \mid \text{HCl}\ (c_2) \mid \text{H}_2, \text{Pt}. \qquad (3.103)$$

The vertical lines in the diagram denote phase boundaries between solid and liquid.

To complete the circuit, so that current can flow, the two end electrodes must of course be connected by wires.

The symbol Pt, H_2 denotes an electrode of spongy platinum or palladium, with hydrogen gas bubbling through it and the immediately adjoining solution at a defined pressure of H_2 (usually at, or very near, 1 atm). The reaction at this electrode (denoting an electron by e^-) is

$$\tfrac{1}{2}H_2 \rightleftharpoons H^+ + e^-.$$

If this reaction runs from left to right, the released electron flows out through the wire into the external circuit; if the reaction goes the other way, the electron comes from the wire, reacts with a (hydrated) proton in the solution, and forms neutral hydrogen.

At the boundary between the solution and the other electrode, made of silver and silver chloride, the reaction involves the formation or disappearance of chloride ions in the solution:

$$Ag + Cl^- \rightleftharpoons AgCl + e^-.$$

If $c_1 > c_2$, the chemical potential of HCl is higher in the cell on the left, and the spontaneous tendency of current flow will be in the direction that causes HCl to be transferred from the left-hand to the right-hand compartment.

Thus, in the left-hand compartment the net reaction will be

$$(H^+ + Cl^-)_{c_1} + Ag \rightarrow AgCl + \tfrac{1}{2}H_2$$

and in the right-hand compartment it will be the reverse of this. The net result is the transfer of HCl from concentration c_1 to c_2, while there is a counter-clockwise flow of electrons in the external circuit; hence a clockwise flow of positive electricity in this circuit. Since the cell is balanced in a potentiometer, the electrical work, per mole of H^+ and Cl^- ions transferred, is **FE**, where **F** is the Faraday equivalent, 96 485 C mol^{-1}. The chemical potential work involved in transferring 1 mol of H^+ and 1 mol of Cl^- ions from concentration c_1 to c_2 is

$$(\mu_{HCl})_2 - (\mu_{HCl})_1 = RT \ln [(a_{HCl})_2/(a_{HCl})_1]. \qquad (3.104)$$

Since HCl is essentially completely ionized in dilute solution, it is natural to resolve μ_{HCl} into the sum of two ionic potentials μ_{H^+} and μ_{Cl^-}, and take the activity of each ion as the product of its concentration by an activity coefficient:

$$a_{HCl} = a_{H^+}a_{Cl^-} = [H^+][Cl^-]\,\gamma_H\,\gamma_{Cl} = [H^+][Cl^-]\,\gamma_\pm^2.$$

The electrical work expended must be equal and opposite to the chemical potential work of ion transfer, under reversible conditions:

$$\mathbf{F}\mathbf{E} = RT\ln\frac{[\mathrm{H}^+]_1[\mathrm{Cl}^-]_1(\gamma_\pm)_1^2}{[\mathrm{H}^+]_2[\mathrm{Cl}^-]_2(\gamma_\pm)_2^2}. \tag{3.105}$$

Since $[\mathrm{H}^+] = [\mathrm{Cl}^-] = c_1$ or c_2, this can be written, after dividing by $\mathbf{F}$, and taking T as 298.15 K, as

$$\mathbf{E} = RT\ \ln\frac{c_1^2(\gamma_\pm)_1^2}{c_2^2(\gamma_\pm)_2^2} = 0.059\ 15\ \log\frac{c_1^2(\gamma_\pm)_1^2}{c_2^2(\gamma_\pm)^2} = 0.1183\ \log\frac{c_1(\gamma_\pm)_1}{c_2(\gamma_\pm)_2}. \tag{3.106}$$

Here the factor 0.059 15 = 2.3026 × 8.314 × 298.15/96 485.

If, for example, c_1 were 10 times c_2, and if the activity coefficient term were about the same in both, the reversible e.m.f. would be 0.1183 V. In pure hydrochloric acid there would be significant error in making such an assumption about the activity coefficient; in dilute solution we could use the Debye–Hückel equation (3.88) to calculate the log $\gamma_\pm$ terms. In biochemical systems, however, we work commonly in the presence of a medium containing neutral salts at a fairly constant ionic strength, so that the log $\gamma_\pm$ terms can be omitted, as a first approximation, in work of moderate accuracy. For a far more detailed and rigorous discussion of electrolyte solutions see Harned and Owen (1958) or Robinson and Stokes (1970). Edsall and Wyman (1958, pp. 429–441) give a rather brief discussion, but with some details that we do not present here.

The double cell described above is conceptually convenient for understanding the relation between the e.m.f. and the chemical work of transferring the ions. However, in practice it is more convenient, and more accurate, to use a cell with a single compartment for the solution between the electrodes:

$\mathrm{Pt,H_2} \mid \mathrm{HCl}\,(c) \mid \mathrm{AgCl,\ Ag}.$

We can measure $\mathbf{E}$ for this cell, first with HCl at concentration c_1, and then at concentration c_2. The difference between the two $\mathbf{E}$ values will be the same as the total e.m.f. of the double cell. The equation for $\mathbf{E}$ of the single cell is

$$\begin{aligned}\mathbf{E} &= \mathbf{E}_0 - \frac{RT}{F}\ln\,(c^2\gamma_\pm^2)\\ &= \mathbf{E}_0 - 0.1183\log c - 0.1183\log\gamma_\pm.\end{aligned} \tag{3.107}$$

Here $\mathbf{E}_0$ is a standard potential corresponding to the value of $\mathbf{E}$ when the HCl is at unit activity. If $c = 0.1$ M, $\mathbf{E} = +0.3524$ V. The positive value for $\mathbf{E}$ indicates that the spontaneous flow of positive current through the cell is from left to right. (In the external circuit, of course, it is in the opposite direction.) If we were to write the diagram of the cell in reverse, with Ag, AgCl on the left and $\mathrm{H_2}$,Pt on the right, then by convention $\mathbf{E}$ would be −0.3524 V. One can determine $\mathbf{E}_0$ by plotting $\mathbf{E} + 0.1183 \log c$ against $c^{1/2}$, and extrapolating to zero ionic strength, since $-\log \gamma_\pm$ is proportional to $c^{1/2}$ at low ionic strengths. Details of procedure and calculation are discussed by Harned and Owen (1958) and Robinson and Stokes (1970). It is of course essential, in drawing conclusions from such electrical studies, to prove that the assumed chemical reactions actually account quantitatively for what happens when the current flows. For the cell we have been discussing, this has been abundantly demonstrated.

One can adjust a cell of this sort so as to be responsive only to changes in the hydrogen ion concentration (or activity) by adding a large excess of a neutral chloride, such as NaCl or KCl. Then, if the chloride ion concentration and the total ionic strength are held constant, say somewhere in the range from 0.15 to 0.25 M, while the HCl concentration is varied, say in the range from 10^{-2} to 10^{-4} M, the terms in $[Cl^-]$ and in $\gamma_\pm$ in (3.105) should cancel out, and the $\mathbf{E}$ value of such a double cell as (3.103) should be given simply by the equation

$$\mathbf{E} = 0.059\,15 \log([H^+]_1/[H^+]_2); \qquad (3.108)$$

that is, the measurement should give directly the relative H^+ ion *concentrations* in the two solutions.

3.23 Calibration and measurement of pH

The type of cell which we have just discussed is in principle very close to the systems used for measurement of pH. There are two important differences:

(1) Cells used for determination of pH contain what is known as a liquid junction, and a calomel (Hg_2Cl_2) electrode is generally used instead of the silver–silver chloride electrode. To protect the solution whose pH is being measured from possible contamination by the mercury compound, a solution of concentrated KCl is interposed between the calomel electrode and the solution under study. The cell is then

$$\text{Pt, } H_2 \mid \text{solution X} \parallel \text{concentrated KCl} \mid Hg_2Cl_2\text{, Hg.} \qquad (3.109)$$

Here the double vertical line denotes the boundary between two liquids of different composition. At such a boundary there is an electrical potential, the liquid junction potential. Its magnitude depends on the concentration and mobility of all the ions in the two solutions adjoining the boundary. There is a large literature on the subject (see the references already cited and also Bates, 1973). The advantage of using concentrated KCl at the boundary is that the mobilities of the K^+ and Cl^- ions in an electric field are nearly equal, and this greatly diminishes the liquid junction potential. Moreover the K^+ and Cl^- ions are present at much higher concentration than any of the ions in the unknown solution (X); hence the liquid junction potential, whatever its value may be, is primarily determined by the concentrated KCl, and is little affected by moderate changes in the composition of solution X. We calibrate the system by measuring **E** when solution X is one of several standard buffers of known pH; we generally choose a standard whose pH is not very far from that of the solution under study. Then the pH of the latter is simply given (at 25 °C) by the equation

$$\mathbf{E} - \mathbf{E}_{\text{standard}} = 0.0591\,(\text{pH}_{\text{X}} - \text{pH}_{\text{standard}}) \tag{3.110}$$

(2) The other special feature of modern apparatus for pH determination is the replacement of the platinum–hydrogen electrode by a thin membrane of a suitable soft glass of a specially defined composition and moderately high conductivity. Such a glass, with solution X on one side of the membrane and a solution of constant composition (e.g. 0.1 M HCl) on the other, responds to changes in the pH of the solution just like a hydrogen electrode, from pH values of 1 or less to quite high pH values, around 12. The glass electrode represents a vast technical simplification; moreover, by eliminating the need for bubbling hydrogen into the liquid it makes possible pH studies of liquids containing various gases, such as oxyhaemoglobin solutions, which could not be studied in the presence of flowing hydrogen gas. (The hydrogen in some systems would of course act as a reducing agent, which is a further disadvantage.)

When S. P. L. Sörensen first defined the concept of pH in 1909, and showed how to measure it, he defined it as the negative logarithm of the H^+ ion concentration. Later several workers proposed the definition of pH as $-\log a_{H^+}$ where a_{H^+} is the hydrogen ion activity. However, there are fundamental difficulties in defining, and therefore in measuring, the activity of an individual ion. The pH, as measured, is not exactly equal either to $-\log\,[H^+]$ or $-\log a_{H^+}$, but the pH concept is based solidly on reproducible experimental operations and is of course indispensable as a guide to the acid–base equilibrium of biological and chemical systems. For further discussion of the theory and practice of pH measurement, see Bates (1973).

3.24 Oxidation–reduction potentials

In general, oxidation–reduction reactions correspond to the equation

$$\text{oxidant} + ne^- \rightleftharpoons \text{reductant}.$$

The value of n is generally either 1 or 2. There are, however, special cases, such as that of haemoglobin, with its four iron atoms, each encased within a planar haem group and surrounded by hydrophobic side chains of one of the α or β protein subunits. The oxidation of ferrous to ferric haem iron, which converts deoxyhaemoglobin to methaemoglobin, occurs in steps, but the process is cooperative (at least under some common circumstances: see Chapter 5 for a discussion of such processes); if some of the haem iron atoms are oxidized, it increases the chance that others will be too. However, haemoglobin is a rather special case, and we confine the rest of our discussion to systems where $n = 1$ or 2.

In many cases, especially with organic compounds, proton transfer accompanies the electron transfer. Often we could write the reaction either in terms of combined electron and proton transfer or in terms of transfer of H atoms. Thus

$$\text{fumarate}^{2-} + 2H^+ + 2e^- \rightleftharpoons \text{succinate}^{2-}$$

or

$$\text{fumarate}^{2-} + 2H \rightleftharpoons \text{succinate}^{2-}.$$

From the point of view of thermodynamics it does not matter which formulation we adopt; indeed other ways of writing the overall process would be possible. In formulating oxidation–reduction potentials, however, it is generally more convenient to write the reaction in terms of combined transfer of protons and electrons. In doing this we are saying nothing about the detailed mechanism of the reaction, of which for the most part we know little.

Table 3.10 lists a number of oxidants of biochemical significance, and their conjugate reductants. We now consider the significance of the oxidation–reduction potentials listed in this table. To measure such a potential one generally employs a cell of the type

$$\text{metal} \,|\, \begin{pmatrix} \text{redox} \\ \text{solution} \end{pmatrix} \,||\, \text{KCl} \,|\, \text{Hg}_2\text{Cl}_2,\ \text{Hg}.$$

The metal electrode, unlike the metal in the hydrogen electrode, is generally a strip of clean, pure, solid noble metal, such as gold or platinum. The ultimate reference standard is a hydrogen electrode half-cell, with hydrogen gas at 1 atm pressure, and H^+ ion in solution at a pH of 0 (unit H^+ activity). The potential of this half-cell is taken as 0 V, by definition, at all temperatures. (It is impossible to measure the potential of an isolated half-cell.) In practice it is convenient to use a calomel half-cell, and calculate the potential with reference to the standard hydrogen electrode by difference, since this difference is experimentally well established.

If we vary the ratio of oxidant [Ox] to reductant [Red] in the redox solution, at constant pH, the measured e.m.f. at 25 °C will vary according to the relation

$$\mathbf{E}_h = \mathbf{E}_0 + \frac{RT}{n\mathbf{F}} \ln \frac{[\text{Ox}]}{[\text{Red}]} = \mathbf{E}_0 + \frac{0.0591}{n} \log \frac{[\text{Ox}]}{[\text{Red}]} \cdot \tag{3.111}$$

Thus $\mathbf{E}_0$ for a particular system is the value of $\mathbf{E}_h$ (relative to the standard hydrogen electrode) when [Ox] = [Red] in that system. For example, if [Ox] is $[Fe^{3+}]$ and [Red] is $[Fe^{2+}]$ (solutions of an inorganic iron salt) then $\mathbf{E}_0$ turns out to be +0.771 V, relative to the standard hydrogen electrode, as listed in Table 3.10.

Table 3.10 Some oxidant–reductant pairs, with their mid-point potentials at pH 7

Oxidant		Reductant	n	$\mathbf{E}_m$ (pH 7)
$H^+ + e^-$	$\rightleftharpoons$	$\frac{1}{2} H_2$	1	−0.421
$NAD^+ + 2H^+ + 2e^-$	$\rightleftharpoons$	$NADH + H^+$	2	−0.320
$NADP^+ + 2H^+ + 2e^-$	$\rightleftharpoons$	$NADPH + H^+$	2	−0.324
Acetaldehyde $+ 2H^+ + 2e^-$	$\rightleftharpoons$	Ethanol	2	−0.197
Pyruvate$^-$ $+ 2H^+ + 2e^-$	$\rightleftharpoons$	Lactate$^-$	2	−0.185
Fumarate^{2-} $+ 2H^+ + 2e^-$	$\rightleftharpoons$	Succinate^{2-}	2	−0.031
Cytochrome (Fe^{3+}) $+ e^-$	$\rightleftharpoons$	Cytochrome *b* (Fe^{2+})	1	+0.030
Cytochrome *c* (Fe^{3+}) $+ e^-$	$\rightleftharpoons$	Cytochrome *c* (Fe^{2+})	1	+0.254
Cytochrome a_3 (Fe^{3+}) $+ e^-$	$\rightleftharpoons$	Cytochrome a_3 (Fe^{2+})	1	+0.385
Fe^{3+} (iron salt) $+ e^-$	$\rightleftharpoons$	Fe^{2+}	1	+0.771
$\frac{1}{2}O_2 + 2H^+ + 2e^-$	$\rightleftharpoons$	H_2O	2	+0.816

The $\mathbf{E}_m$ values are taken from Lehninger (1975, p. 479). For a detailed critical discussion of $\mathbf{E}_m$ values obtained by various workers, see Clark (1960).

Since biochemists generally work at pH values near 7, it is more convenient for them to express their data in terms of redox equilibria at pH 7, or sometimes at another neighbouring pH value. The potential of the hydrogen electrode in conjunction with a solution at pH 7 is given (to two places of decimals) as −0.42 V. This follows from the fact that the potential becomes more negative by approximately 0.06 V for each unit increase in pH from 0 to 7. The numerical factor in (3.111) becomes 0.0601 at 30 °C, and for simplicity we write it simply as 0.06 from now on. Following Clark (1960) we denote the mid-point potential of a redox system at pH 7, referred to the standard hydrogen electrode, as $\mathbf{E}_m$, or $\mathbf{E}_{m7}$ if we wish to emphasize the pH:

$$\mathbf{E}_h = \mathbf{E}_{m7} + \frac{0.06}{n} \log \frac{[\mathrm{Ox}]}{[\mathrm{Red}]} \cdot \qquad (3.112)$$

If $n = 2$, the numerical factor in (3.112) becomes 0.03.

The value of $\mathbf{E}_m$ for a particular system depends on the pH, if protons are given up or taken on, along with the electron transfer. This may occur as a necessary accompaniment of the electron transfer, as in the ethanol–acetaldehyde system (see Table 3.10), in which $\mathbf{E}_m$ becomes more negative by 0.06 V for each increase of one unit in pH. Other situations arise in which there are acidic groups in oxidant or reductant or both, which may be stronger (lower pK values) in one than in the other. If no proton transfer accompanies the electron transfer in a particular region of pH, the $\mathbf{E}_m$ value of the system is independent of pH in that region. For a detailed discussion of the relations involved, see Clark (1960, Chapter 4).

It is not always easy to establish equilibrium in the case of an oxidation–reduction system. In many cases of biochemical interest the e.m.f. responds sluggishly or erratically. Often equilibrium is not established unless an enzyme is present to catalyse the process. The first use of an enzyme to make possible the measurement of a redox system was by Lehmann (1930) and Borsook and Schott (1931), who used succinic dehydrogenase to catalyse electron transfer in the succinate–fumarate system.

The importance of these $\mathbf{E}_m$ values is that they serve to define the direction in which the reaction can run when two different oxidant–reductant pairs are together in solution; as in thermodynamics generally, processes that are thermodynamically possible, with a negative ΔG value, may not occur at an appreciable rate. In biochemical systems the presence of an enzyme is often essential, as indicated above. This is in fact a safeguard to the organism. If such processes occurred too readily, without a catalyst, they could uselessly dissipate precious Gibbs energy.

The oxidant of the system of higher (more positive) potential tends to oxidize the reductant of the system of lower potential. For instance, consider what happens when the components of the NAD^+–NADH system ($\mathbf{E}_{m1} = -0.320$ V) come into equilibrium with the acetaldehyde–ethanol system ($\mathbf{E}_{m2} = -0.197$ V) at pH 7, in the presence of alcohol dehydrogenase to catalyse the electron transfer and a buffer to keep the pH relatively constant. The potential at equilibrium, $\mathbf{E}_h$, must satisfy two equations:

$$\mathbf{E}_h = \mathbf{E}_{m1} + 0.03 \log \frac{[NAD^+]}{[NADH]} = \mathbf{E}_{m2} + 0.03 \log \frac{[CH_3CHO]}{[C_2H_5OH]} .$$

Therefore

$$\mathbf{E}_{m2} - \mathbf{E}_{m1} = -0.197 - (-0.320) = 0.123 = 0.03 \log \frac{[NAD^+][C_2H_5OH]}{[NADH][CH_3CHO]}$$

or

$$\frac{[NAD^+][C_2H_5OH]}{[NADH][CH_3CHO]} = 12.6 \times 10^3.$$

Thus if $[NAD^+] = [NADH]$ at equilibrium, the alcohol–aldehyde system is almost completely converted to ethanol; in other words, at pH 7 the reaction

$$CH_3CHO + NADH + H^+ \rightharpoonup C_2H_5OH + NAD^+$$

goes almost comlpetely from left to right, acetaldehyde being an oxidant of higher potential than NAD^+. For this example values from Table 3.10 were used for the calculation of the equilibrium at pH 7 and 30 °C. For a more recent evaluation of the thermodynamic functions of this reaction see Section 4.3.

We note that a hydrogen ion is involved in the reaction, though it is not included in the equilibrium constant as written above. If the reaction occurs in a well buffered medium, $[H^+]$ remains essentially constant, and the equilibrium constant, with the $[H^+]$ term omitted, is a constant *for this particular pH value*. It will, however, change if the pH changes. Such 'apparent' equilibrium constants, which are constant only for specified pH values, are considered further in Chapter 4.

If two oxidant–reductant pairs, involving transfer of only a single electron, were brought together in a similar fashion, the logarithmic term would be multiplied by 0.06 instead of 0.03. Consider cytochrome *c* and cytochrome a_3 with $\mathbf{E}_m$ values 0.254 V and 0.385 V respectively. If equilibrium were attained, we would have

$$\mathbf{E}_h = \mathbf{E}_m(\text{cyt } c) + 0.06 \log \frac{[\text{cyt } c^{3+}]}{[\text{cyt } c^{2+}]} = \mathbf{E}_m(\text{cyt } a_3) + 0.06 \log \frac{[\text{cyt } {a_3}^{3+}]}{[\text{cyt } {a_3}^{2+}]} .$$

Thus

$$\mathbf{E}_m(\text{cyt } a_3) - \mathbf{E}_m(\text{cyt } c) = 0.131 = 0.06 \log \frac{[\text{cyt } c^{3+}][\text{cyt } {a_3}^{2+}]}{[\text{cyt } c^{2+}][\text{cyt } {a_3}^{3+}]}$$

Hence

$$\frac{[\text{cyt } c^{3+}][\text{cyt } a_3{}^{2+}]}{[\text{cyt } c^{2+}][\text{cyt } a_3{}^{3+}]} = \text{antilog}\,\frac{0.131}{0.06} = 153\ .$$

Thus, although the potential difference betwen the two cytochrome systems is slightly greater than that between the NAD^+ and the acetaldehyde system, the oxidation of the reductant of the system of lower potential by the oxidant of higher potential does not proceed so far at equilibrium, in a one-electron transfer.

In the living organism, however, these equilibrium considerations are important primarily because they indicate the direction of electron transfer, from systems of low potential such as NAD^+ and $NADP^+$, to systems of higher potential such as the aldehyde–alcohol or the pyruvate–lactate systems, or ubiquinone and the flavins, and on to the cytochromes, cytochrome oxidase, and molecular oxygen. It is the dynamic system of electron flow from low potential to high potential systems, in a steady state, that characterizes the living organism; nevertheless, thermodynamic data give us essential information that helps us to visualize what is going on in that steady state.

It is a striking fact that, in systems involving two-electron transfer, the two electrons commonly go on and off simultaneously, in what looks like a completely cooperative process. This is by no means always so; there are many systems in which the electron transfer proceeds in two distinct steps. The intermediate in such a process is necessarily a free radical, with one unpaired electron, and is therefore paramagnetic. Such intermediates were first reported in 1931, independently by L. Michaelis and B. Elema, and Michaelis christened them semiquinones. They are important in many biochemical systems, for example in the flavin enzymes. They are well discussed by Clark (1960, Chapter 7) and in a number of publications by Michaelis (e.g. Michaelis, 1940).

The application of the thermodynamics of oxidation–reduction potentials to problems in bioenergetics is treated in detail by Walz (1979). In Walz's review one finds a good summary of the compromises which have to be made when thermodynamic rigour conflicts with the interpretation of complex biological systems.

3.25 Electrostatic fields and electrochemical potentials

Up to this point we have generally confined our discussion to systems that could be adequately described in terms of pressure, temperature, and the composition of the components. There are, however, important systems for which it is imperative to consider the influence of other variables, such as electrostatic fields or gravitational or centrifugal fields. In seeking to understand the properties of ions, for instance, we must enlarge our concept of chemical potential to include the influence of electrostatic fields.

A charged body, of charge q coulombs, which can move between two phases with different electrostatic potentials, acquires or loses electrical energy equal to $q\Delta\Phi$ on moving from one phase to the other, where Φ is the electrical potential difference measured in volts. For a mole of ions of valence Z_i (for instance $Z = +2$ for Ca^{2+}, -1 for Cl^-) this electrical energy is equal to $Z_i\mathbf{F}\Phi$. The total molar potential of the ions must include the usual concentration (or activity) term for the chemical potential, with this added electrical term. The resultant combined expression, denoted by the symbol $\bar{\mu}_i$, is called the electrochemical potential of the ion:

$$\bar{\mu}_i = \mu_i^\circ + RT \ln c_i + Z_i\mathbf{F}\Delta\Phi. \tag{3.113}$$

If all the ions in the system can move freely between phases, $\Delta\Phi$ becomes zero, and the electrochemical potential is reduced to the ordinary chemical potential. However, if there are constraints on the movements of some of the ions, $\Delta\Phi$ will not in general be zero and we must consider all the terms in equation (3.113).

One relatively simple case arises if there are two solutions, each containing the same two ions of a salt, separated by a membrane permeable to only one of the ions, the concentration of salt being different on the two sides. Consider a membrane separating two K^+ Cl^- solutions at concentrations c_1 and c_2 ($c_1 > c_2$) and suppose the membrane to be permeable only to K^+ ions. The c_1 solution is on the right. Then a very small net number of K^+ ions will cross the membrane from right to left, but Cl^- ions cannot cross with them. The resulting excess of K^+ over Cl^- ions on the left will be exceedingly small, below analytical detection, but it will make the electrostatic potential higher (more positive) on the left than on the right. At equilibrium, however, the electrochemical potential of the potassium ions must be the same on both sides:

$$(\bar{\mu}_i)_1 = (\bar{\mu}_i)_2 = \mu_i^\circ + RT \ln(c_2) + Z_i\mathbf{F}\Phi_2 = \mu_i^\circ + RT \ln(c_1) + Z_i\mathbf{F}\Phi_1.$$

Hence, for the potential difference,

$$\Delta\Phi = \Phi_1 - \Phi_2 = \frac{RT}{Z_i\mathbf{F}} \ln\left(\frac{c_2}{c_1}\right) = 0.06 \log\left(\frac{c_2}{c_1}\right) \cdot \tag{3.114}$$

Thus, since $c_2 < c_1$, the right-hand term in (3.114) is negative, and $\Phi_2 > \Phi_1$. This means that the K^+ ion electrostatic potential is higher on the side where its concentration is lower. The K^+ ion moves, as would be expected in view of its concentration gradient, from c_1 to c_2; but as the Cl^- ion cannot cross the membrane, the positive electrostatic potential, thus built up, brings the system to equilibrium when the excess of K^+ ions over Cl^- in phase 2 is still (by analysis) undetectably small.

We have already seen an uneven distribution of ions between two phases, when a charged macromolecule is present on one side of a membrane, along with diffusible anions and cations that can pass the membrane freely (see equations (3.100) and (3.100a)). Consider (say) a positively charged protein inside the membrane, with sodium and chloride ions present in both the inner and outer solutions. Then we know that the concentration of both those ions must be different on the two sides. There will be more chloride than sodium on the inside, to balance the charge on the protein, and the ratios of the two ions (inside to outside) must obey equation (3.100a):

$$\frac{[\mathrm{Cl}^-]_\mathrm{i}}{[\mathrm{Cl}^-]_\mathrm{o}} = \frac{[\mathrm{Na}^+]_\mathrm{o}}{[\mathrm{Na}^+]_\mathrm{i}} > 1. \tag{3.100a}$$

By the same reasoning which we have already given, this requires the existence of an electrical potential difference – the Donnan potential – between the two phases.

Consider the electrochemical potential of the sodium ion on the inside and outside, remembering that the inside is under a higher pressure – the osmotic pressure Π – than the outside, which gives a term ΠV_{Na} in the equation for the total potential. The electrochemical potential must be the same on both sides at equilibrium, and $Z_{\mathrm{Na}} = +1$. Using the subscripts 'in' and 'out' for the two solutions,

$$\begin{aligned} \bar{\mu}_{\mathrm{Na}} - \mu^\circ_{\mathrm{Na}} &= RT\ln(c_{\mathrm{Na}})_{\mathrm{in}} + \Pi\, V_{\mathrm{Na}} + \mathbf{F}\Phi_{\mathrm{in}} \\ &= RT\ln(c_{\mathrm{Na}})_{\mathrm{out}} + \mathbf{F}\Phi_{\mathrm{out}}. \end{aligned} \tag{3.115}$$

The term ΠV_{Na} turns out to be relatively small, and we can omit it. Then, on rearranging (3.115) and inserting the numerical factor 0.06, we have

$$\Delta\Phi = \Phi_{\mathrm{in}} - \Phi_{\mathrm{out}} = \frac{RT}{\mathbf{F}}\ln\frac{(c_{\mathrm{Na}})_{\mathrm{out}}}{(c_{\mathrm{Na}})_{\mathrm{in}}} = 0.06\log\frac{(c_{\mathrm{Na}})_{\mathrm{out}}}{(c_{\mathrm{Na}})_{\mathrm{in}}}\,. \tag{3.116}$$

Since $(c_{\mathrm{Na}})_{\mathrm{in}} < (c_{\mathrm{Na}})_{\mathrm{out}}$, this means that the potential $\Delta\Phi$ is positive; the electrostatic potential is higher inside than outside. On applying the same reasoning to the distribution of chloride ions, remembering that $Z_i = -1$ for chloride, we get the same result for the electrostatic potential:

$$\Delta\Phi = -\frac{RT}{\mathbf{F}}\ln\frac{[\mathrm{Cl}]_{\mathrm{out}}}{[\mathrm{Cl}]_{\mathrm{in}}} = \frac{RT}{\mathbf{F}}\ln\frac{[\mathrm{Cl}]_{\mathrm{in}}}{[\mathrm{Cl}]_{\mathrm{out}}}\,. \tag{3.117}$$

Thus, if the macro-ion inside the membrane is positively charged, the inner solution is positive, relative to the outside. If the charge on the macro-ion is negative, the inner solution is negative relative to the outer.

3.26 Electrochemical potentials in physiology: resting potentials and action potentials in nerve

Electrochemical potentials are important in physiological systems. We consider briefly the situation inside and outside a nerve axon. We start with a hypothetical axon permeable only to K^+ and Cl^- and the following conditions:

	Interior conc./mM	Exterior conc./mM
K^+	400	20
Na^+	50	440
Cl^-	120	550
Non-diffusible organic anions	350	–

Potassium ions will diffuse out of the axon along the concentration gradient. This diffusion results in a potential across the membrane (negative inside, since the non-diffusible ions inside are negative) and will – in this hypothetical axon – reach an equilibrium when the electrochemical potential of K^+ ions is equal on both sides. If the ratio $[K^+]_{in}/[K^+]_{out} = 20$, as above, then the membrane potential $\Delta\Phi$ should be 0.06 log 20 = 0.078 V, with the inside negative.

A more realistic description of the situation in nerve axons requires consideration of steady states and transitions between them. The resting nerve has a limited permeability to sodium ions, but an enzyme is present which pumps Na^+ back into the axon, using the hydrolysis of ATP as an energy source (see Chapter 4). To describe this system we require an extended equation (the Goldman equation) to take account of membrane permeability to the ions. The membrane potential is described by

$$\mathbf{E} = \frac{RT}{n\mathbf{F}} \ln \left[\frac{[K^+]_{external} P_{K^+} + [Na^+]_{external} P_{Na^+}}{[K^+]_{internal} P_{K^+} + [Na^+]_{internal} P_{Na^+}} \right], \tag{3.118}$$

where P_{K^+} and P_{Na^+} are the relative permeabilities of the membranes to these ions. Clearly, non-diffusible ions do not take part in equilibration, and ions with restricted diffusion do so in proportion to their permeation. In the resting nerve $P_{K^+} = 100\ P_{Na^+}$. If the logarithmic term is divided by P_{Na^+} and evaluated,

$$\mathbf{E} = \frac{8.314 \times 310}{96\,485} \ln \left(\frac{20 \times 100 + 440}{400 \times 100 + 50} \right) = -75 \text{ mV},$$

which is the resting potential; a figure very close to that calculated above.

During an action potential P_{Na^+}/P_{K^+} changes to $\simeq 12$ and the resulting membrane potential is +55 mV. The change of +130 mV is due to the inward flow of sodium ions as the membrane reverses its relative permeability to cations. It is interesting to calculate the actual change in ion concentration during this change in potential. The capacitance of the nerve membranes is 1 μF cm^{-2} and the charge Q = capacitance × voltage = 10^{-6} F cm^{-2} × 0.13 V, which is 1.3×10^{-7} C cm^{-2} or a movement of 1.3×10^{-12} mol cm^{-2}. In large nerve fibres this corresponds to less than 1 part in 10^6 of the total ions. A good discussion of the elementary theory of energy storage in capacitors is found in Katz (1966) and in Kuffler and Nichols (1976). The treatment given in these references is particularly relevant to biological systems and to measuring devices used for their study.

3.27 Equilibrium in the ultracentrifuge

In a centrifugal field, the potential of molecules in solution depends not only on pressure, temperature, and composition, but also on the strength of the centrifugal field. We denote the molar mass of molecule i by M_i. The force acting upon 1 mol of molecules is equal to M_i times the centrifugal acceleration; the latter is equal to $\omega^2 r$, where ω is the angular velocity of the centrifuge in radians per second, and r is the radial distance from the centre of rotation. Thus the increment in centrifugal energy of the molecule of mass m_i, when it moves through a distance increment dr, is $-m_i\omega^2 r\,\mathrm{d}r$, and the centrifugal force *per mole* is $M_i\omega^2 r$, where M_i is the molar mass. The sign of this term is negative, since the applied field tends to drive the molecules outward. In considering the total variation in chemical potential of the molecule we must take account of this form of its energy along with pressure, temperature, and composition. Thus for the total increment in μ_i we write

$$\mathrm{d}\mu_i = \left(\frac{\partial \mu_i}{\partial T}\right)\mathrm{d}T + \left(\frac{\partial \mu_i}{\partial p}\right)\mathrm{d}p + \left(\frac{\partial \mu_i}{\partial c_i}\right)\mathrm{d}c_i - M_i\omega^2 r\,\mathrm{d}r. \tag{3.119}$$

The first term on the right is zero, since it is experimentally possible to maintain a constant temperature throughout the rotating cell. The second term $(\partial\mu_i/\partial p)$ is the partial molar volume (V_i) as before, and p is the hydrostatic pressure in the liquid that develops as a result of the centrifugal field. The increment in p, as r increases, in a layer of liquid of unit cross-section and thickness dr is

$$\mathrm{d}p = \rho\,\omega^2 r\,\mathrm{d}r,$$

where ρ is the density of the liquid at that point. The term $(\partial\mu_i/\partial c_i)\,\mathrm{d}c_i$, where we have assumed that we may set $a_i = c_i$, becomes

$$\left(\frac{\partial \mu_i}{\partial c_i}\right)\mathrm{d}c_i = RT\,\mathrm{d}(\ln c_i).$$

At equilibrium the *total* potential of component i must be the same throughout the cell. Hence we can rewrite equation (3.119), setting $\mathrm{d}\mu_i = 0$, as

$$\mathrm{d}\mu_i = RT\,\mathrm{d}(\ln c_i) + (V_i\rho - M_i)\omega^2 r\,\mathrm{d}r = 0. \tag{3.120}$$

Moreover, we can set the partial molar volume V_i equal to $v_i M_i$, where v_i is the partial *specific* volume. Then on rearranging terms we have

$$RT\,\mathrm{d}(\ln c_i) = M_i(1 - v_i\rho)\omega^2 r\,\mathrm{d}r$$

or, since M_i is the quantity we seek to measure, we can write

$$M_i = \frac{RT\,\mathrm{d}(\ln c_i)}{(1 - v_i\rho)\omega^2 r\,\mathrm{d}r}\,. \tag{3.121}$$

Integration of this equation between two distances r_1 and r_2 from the centre of rotation, and the corresponding concentrations, c_{i1} and c_{i2}, at those distances gives, after rearranging terms, the equation for the molar mass M_i:

$$M_i = \frac{2\,RT\,\ln(c_{i2}/c_{i1})}{\omega^2(1 - v_i\rho)(r_2^2 - r_1^2)}\,. \tag{3.122}$$

Note that we have assumed here that v_i and ρ are constant throughout the cell, even though the pressure increases as we go outward from the centre of rotation. This assumption is justified for practical purposes because of the low compressibility of water, so the density changes very little between the inner and outer ends of the cell.

Here we note that M_i is the molar mass of component i, commonly expressed in grams per mole; in SI units, however, it is in kilograms per mole. It is numerically equal to the 'molecular weight', but the latter quantity, as officially defined by IUPAC, is a pure number, being the ratio of the mass of the molecule in question to one-twelfth of the mass of one atom of the carbon isotope of mass 12 (^{12}C). To emphasize the nature of 'molecular weight' (properly 'relative molecular mass') as a ratio, it is now customary to denote it by the symbol M_r. In equation (3.122) above, however, M_i is actually a mass, as we can see that it must be by considering the terms on the right-hand side of the equation. RT has the dimensions of energy (ml^2t^{-2}); the logarithmic term is of course dimensionless, as is the term $(1 - v_i\rho)$ in the denominator (v_i is a reciprocal density). The angular velocity ω has the dimensions of reciprocal time (an angle is dimensionless in terms of length); so ω^2 has the dimension t^{-2}; and the r^2 terms are of course of dimension l^2. Hence the whole term on the right has the dimension of mass: $ml^2t^{-2}/l^2t^{-2} = m$; and M_i on the left must be a mass. This exemplifies the use of dimensional analysis in checking equations, to determine that they are dimensionally consistent.

Equation (3.122) is the simplest form of the equation for equilibrium of a macromolecule in the ultracentrifuge. It assumes tacitly that there are only two components, a pure solvent and a single kind of dissolved macromolecule; and it assumes that we can take the concentration of the latter equal to its activity. Nevertheless, it commonly works quite well when the solvent is not pure water, but a buffered salt solution; and we can often disregard variations in activity coefficients.

The equilibrium equation (3.122) can also be derived in an entirely different way, by considering the balance of forces acting on a solute molecule. The centrifugal force tends to drive the macromolecule outward (if its partial specific volume is less than that of the solvent); this tends to set up a concentration gradient, which produces an opposing force, due to diffusion, which tends to eliminate the gradient. At equilibrium the two forces are in balance throughout the cell, and the result is equation (3.122). For the details of the derivation, see Svedberg and Pedersen (1940) or Schachman (1959).

The commonest use of the ultracentrifuge is for measurement of sedimentation velocities of macromolecules, and the determination of the sedimentation coefficient, defined as $s = (\mathrm{d}r/\mathrm{d}t)/\omega^2 r$. Molar masses can be obtained if the diffusion coefficient (D) is also known from a separate measurement. In that case the molar mass is given, in the simplest case, by the equation

$$M_i = RTs/D(1 - V_i\rho). \qquad (3.123)$$

The values of s and D are functions of concentration, temperature, and the viscosity of the solution (see the references mentioned above; also Williams, 1972). Equation (3.123) cannot be derived from equilibrium thermodynamics, but can be derived from the extension of thermodynamics to irreversible processes, originally developed by Lars Onsager. This lies outside the scope of this book; see for instance Katchalsky and Curran (1965).

Consider a calculation for a specific case of the application of the equilibrium equation (3.122). Suppose we want to get the molar mass of a protein molecule, for which we have a rough estimate that the value is around 30 000 g mol^{-1} (30 kg mol^{-1}). It is convenient for measurement to choose the speed of rotation so that the concentration at the bottom of the cell at equilibrium is (say) four times as great as at the top. The top of the cell is 5 cm (0.05 m), and the bottom is 6 cm (0.06 m) from the centre of rotation. What is the approximate value of ω to choose, so that (c_2/c_1) will be around 4? We note that $\ln 4 = 1.4$; we take the solvent density $\rho = 1$, set $T = 300$ K, and take $v_i = 0.75\ \mathrm{cm}^3\ \mathrm{g}^{-1}$, an average value for proteins. Then for the desired value of ω^2, using SI units (masses in kilograms and distances in metres), we have

$$\omega^2 = \frac{2\,RT\ln(c_2/c_1)}{M_i(1 - v_i\rho)(r_2^2 - r_1^2)} = \frac{2 \times 8.31 \times 300 \times 1.4}{30(0.25)(0.0036 - 0.0025)} = 8.5 \times 10^5$$

or

$$\omega = 920\ \mathrm{s}^{-1}.$$

Now $\omega = 2\pi$ times the number of revolutions per second, or $2\pi/60$ times the number of revolutions per minute. So, to obtain $c_2/c_1 = 4$ we would run the centrifuge at around 8500 rev min^{-1}.

We can use the same line of reasoning to calculate the distribution of concentration of a gas in the atmosphere, in the Earth's gravitational field. For simplicity we can consider the atmosphere as if it were composed of a single gas of molar mass M, and take the temperature as a constant (say 300 K). The gas can also be treated as perfect, and its pressure is RT times its concentration.

Then the total potential of the gas is a function of its pressure (or concentration) and of the gravitational potential energy $\boldsymbol{M}g\boldsymbol{r}$, where r is the vertical distance from the Earth's surface. For the variation of the total potential of the gas, with variation in r, we have, at equilibrium, with μ_i constant at all r values,

$$\mathrm{d}\mu_i = RT\,\mathrm{d}(\ln p_i) + M\boldsymbol{g}\,\mathrm{d}r = 0.$$

Hence, on integration,

$$\ln(p_i/p_{i0}) = -\frac{M\boldsymbol{g}}{RT}(r - r_0), \tag{3.124}$$

where p_{i0} and r_0 are values at the Earth's surface. On taking exponentials,

$$p_i/p_{i0} = \exp[-\ M\boldsymbol{g}(r - r_0)/RT]. \tag{3.125}$$

This is known as the hypsometric law for variation of atmospheric pressure with altitude. It is readily extended to a mixture of gases by setting up a set of equations, identical in form to (3.125), in which p_i/p_{i0} refers to the partial pressure of component i, and M_i is its molar mass.

As an application of (3.125), consider how high we must go in the atmosphere to reach a level where the pressure has fallen to half its value at sea level, i.e. the level at which $p_i/p_{i0} = ½$. Taking logarithms, we have, from (3.125),

$$\ln(½) = -0.69 = -M\boldsymbol{g}(r - r_0)/RT.$$

Suppose we consider the atmosphere as composed of oxygen, with $M = 32\ \mathrm{g\ mol^{-1}} = 0.032\ \mathrm{kg\ mol^{-1}}$; $R = 8.3\ \mathrm{J\ K^{-1}\ mol^{-1}}$. We can take $T = 300$ K and $\boldsymbol{g} = 980\ \mathrm{cm\ s^{-2}} = 9.8\ \mathrm{m\ s^{-2}}$. Then

$$r - r_0 = \frac{0.69 \times 8.3 \times 300}{9.8 \times 0.032} = 5500\ \mathrm{m} = 5.5\ \mathrm{km}.$$

If we had taken the atmosphere as being composed of nitrogen ($M = 0.028\ \mathrm{kg\ mol^{-1}}$), the corresponding height would be 6.3 km. Obviously only a rough calculation is worth making here, since we have made the rather unrealistic assumption that the temperature is the same at all levels, and of course have neglected pressure fluctuations. Nevertheless, it is useful to make such a calculation before setting out to climb a high mountain.

3.28 Thermodynamics of systems subject to mechanical stress: elasticity of rubber and rubber-like systems

Elastic bodies, such as rubber and various synthetic polymers that resemble rubber, have unusual mechanical properties. They can be stretched to several times their initial length, and return very nearly to their original state when the stretching force is removed. In contrast, metals and most other materials lengthen only slightly, even under much larger stretching forces, and are liable to rupture if the amount of stretch is more than 1 or 2% of the initial length.

In 1806 John Gough of Manchester, England, reported some important experiments on rubber. On sudden stretching of a rubber band that touched his lips he noted a sensation of increased warmth in the rubber, which became more marked the greater the extension. On permitting the rubber to shorten, the temperature immediately fell again. In another experiment, he fastened a band of rubber to a horizontal rack, and on suspending a weight from the other end, he noted that the band shortened on heating and lengthened on cooling. With the later development of thermodynamics, it became apparent that each of Gough's two observations implied the other. In 1859 Joule published an extensive study of the thermoelastic properties of various substances – metals, wood, and especially vulcanized rubber. He gave quantitative confirmation to Gough's findings, which had involved unvulcanized rubber. The temperature rise on stretching was small, only about 0.05 K when the rubber was stretched to twice its initial length, but it was reproducible. (If the stretch was very small, around 10% of initial length, there was a very slight cooling instead of heating.)

Kelvin, before Joule's experiments, had already derived the thermodynamic relations involved, and we give them here in modern terminology. First we note that the rapid stretch, with applied force, and the contraction when the force is removed, are reversible processes if the stretch is not too great – say not more than two or three times the initial length. Furthermore, the process is essentially adiabatic; in rapid stretching and shortening the rubber does not have time to exchange appreciable amounts of heat with its surroundings. If a process is both reversible and adiabatic, it must be occurring at constant entropy, since $\mathrm{d}S = \mathrm{d}Q/T$ for a reversible process and here $\mathrm{d}Q$ is zero.

Now we can extend the basic equation for internal energy (U) as a function of entropy and volume (equation (3.13)) by including an additional term for the work done on extending the rubber against a force (f) that tends to retract it. If the increment in length is $\mathrm{d}L$, the elastic work done is equal to $f\mathrm{d}L$. This work is in addition to the pressure–volume work, and of opposite sign. (The term $p\mathrm{d}V$ measures work done *by* the system; the term $f\mathrm{d}L$ measures work done *on* the system.) Thus the extended form of equation (3.13) becomes, on the assumption that the rubber is a single component,

$$\mathrm{d}U = T\mathrm{d}S - p\mathrm{d}V + f\mathrm{d}L. \tag{3.126}$$

Now the change of volume on stretching is very small, and as a good approximation we can set $p\mathrm{d}V = 0$. Then U is a function of S and L, and we can write

$$\mathrm{d}U = T\mathrm{d}S + f\mathrm{d}L = \left(\frac{\partial U}{\partial S}\right)_{V,L} \mathrm{d}S + \left(\frac{\partial U}{\partial L}\right)_{V,S} \mathrm{d}L. \tag{3.127}$$

On cross-differentiation,

$$\frac{\partial^2 U}{\partial S \partial L} = \left(\frac{\partial T}{\partial L}\right)_{S,V} = \left(\frac{\partial f}{\partial S}\right)_{L,V} = \frac{T}{C_p}\left(\frac{\partial f}{\partial T}\right)_{L,V} . \tag{3.128}$$

The last equality follows since $\mathrm{d}S = (C_p/T)\mathrm{d}T$.

This shows the necessary thermodynamic relation between Gough's two experiments. Since the temperature rises on a rapid adiabatic stretch at constant S, $(\partial T/\partial L)_{S,V}$ is positive. From the last term in (3.128) we see that the force of retraction at constant length increases with temperature; thus if the rubber is free to shorten it will do so. Either observation implies the other.† Whereas for rubber the coefficients in (3.128) are positive, in metals and most other materials they are negative. Thus a metal wire becomes slightly cooler on applying a stretching force, and it lengthens on heating at constant stress.

Most recent work on elasticity of rubber and related systems has been done under isothermal rather than adiabatic conditions, and at constant pressure. Thus we consider an expanded form of equation (3.31) for the variation in Gibbs energy G, where the work term $f\mathrm{d}L$ contributes directly to G:

$$\mathrm{d}G = -S\mathrm{d}T + V\mathrm{d}p + f\mathrm{d}L. \tag{3.129}$$

Since the system is at constant pressure, $\mathrm{d}p = 0$, and cross-differentiation as before gives

$$\left(\frac{\partial f}{\partial T}\right)_{p,L} = -\left(\frac{\partial S}{\partial L}\right)_{p,T} . \tag{3.130}$$

Since we already know that the left-hand side of this equation is positive, it immediately follows that $(\partial S/\partial L)_{p,T}$ is negative; the entropy *decreases* as the rubber is extended. The tendency of the stretched rubber to shorten is driven by the spontaneous tendency for entropy to increase.

†What Gough actually observed was a shortening of the rubber at constant load (constant f) with rise of temperature. From (3.128), $(\partial f/\partial T)_L$ is positive for rubber. It is easy to see physically that therefore $(\partial L/\partial T)_f$ must be negative, which is what Gough observed. To see this mathematically, note that f is a function of both L and T:

$$\mathrm{d}f = (\partial f/\partial L)_T\, \mathrm{d}L + (\partial f/\partial T)_L\, \mathrm{d}T.$$

If we impose the condition that f is constant ($\mathrm{d}f = 0$) then, by the same procedure as that used in deriving (3.26) from (3.25), we obtain

$$\left(\frac{\partial L}{\partial T}\right)_f = -\frac{(\partial f/\partial T)_L}{(\partial f/\partial L)_T}$$

Both terms on the right-hand side are positive. We know this already for the numerator, from (3.128). For the denominator, it is obvious that the retractile force f must increase as L becomes greater than L_0, its resting length. Hence, since both terms on the right are positive, $(\partial L/\partial T)_f$ must be negative.

Since $f = (\partial G/\partial L)_{p,T}$, using equation (3.130) and the equation $\mathrm{d}G = \mathrm{d}H - T\mathrm{d}S$, we can write

$$f = \left(\frac{\partial H}{\partial L}\right)_{p,T} - T\left(\frac{\partial S}{\partial L}\right)_{p,T} = \left(\frac{\partial H}{\partial L}\right)_{p,T} + T\left(\frac{\partial f}{\partial T}\right)_{p,L} \tag{3.131}$$

for f as a function of L and T at constant p. This is an equation of state for the force f. Since the system is at constant pressure, and since the change of volume with length is very small indeed, and since $\mathrm{d}U = \mathrm{d}H - \mathrm{d}(pV)$, we can, with little error, take $\mathrm{d}H = \mathrm{d}U$, and the equation of state for f becomes

$$f = \left(\frac{\partial U}{\partial L}\right)_{p,T} + T\left(\frac{\partial f}{\partial T}\right)_{p,L} . \tag{3.132}$$

The first term on the right-hand side gives the change of internal energy with length, and the second depends, from (3.130), on the change of entropy with length. If f turns out to be proportional to the absolute temperature, we can write $(\partial f/\partial T) = B$, a constant, and in that case $f = BT$ over the temperature range of the experiments. Then the internal energy term would drop out, and the elastic force would be entirely due to the decrease of entropy with increasing length. In fact this turns out to be very nearly true, from several careful studies that have been made on the determination of the retractile force f, at constant length, over a range of values of temperature and at various lengths. We must note that, to obtain thoroughly reliable values of the energy and entropy terms in such studies, it is necessary to determine the values at constant volume rather than at constant pressure; that is, we must replace (3.132) by

$$f = \left(\frac{\partial U}{\partial L}\right)_{T,V} + T\left(\frac{\partial f}{\partial T}\right)_{V,L} . \tag{3.133}$$

It would be technically exceedingly difficult – indeed almost impossible – to keep the volume of the rubber actually constant during the measurements, but it can be shown (see Flory, 1953, pp. 444 and 489 ff.) that a good approximation is obtainable from the relation

$$-(\partial S/\partial L)_{T,V} \simeq (\partial f/\partial T)_{p,\alpha}, \tag{3.134}$$

where the elongation $\alpha = L/L_0$. Thus the change in force with temperature can be measured at constant pressure, while varying the length so as to maintain a constant ratio, *at each temperature*, between L and the unstressed length L_0.

In this analysis we have assumed that the entire sequence of events involved in the changes of stress and length is reversible. Actually this is by no means always so. Both natural and synthetic rubbers, on stretching to several times their initial length, tend to 'crystallize', in the sense that the fibrillar elements tend to pack into a new position of more or less parallel side-by-side orientation. When crystallization occurs, there is a decrease of internal energy, such as occurs when any solid phase forms from the corresponding liquid, and the term $(\partial U/\partial L)_{T,V}$ is no longer equal to zero, as it is in the case of an ideal rubber whose behaviour on stretching or shortening is strictly reversible. Commonly it is possible by moderate heating to induce the crystallized rubber to contract again and return to its original state; but the return to that state follows a different path, i.e. the system shows hysteresis.

Finally we note the molecular interpretation of the fact that the entropy of rubber-like substances decreases on elongation at constant temperature. We have already indicated the clue to this phenomenon in the preceding paragraph. Such substances are composed of long-chain polymers, cross-linked in places here and there, but with the orientations of the chains distributed in essentially random fashion. As the chains become extended by applied force, they are pulled out into a more nearly parallel orientation, relative to one another; the disorderly arrangement of the random chains is replaced by partial order, and the entropy decreases. This is in contrast to the situation in a metal wire, where the degree of internal order is little altered by a stretching force, but the internal energy is markedly affected.

These considerations of course have biological significance. Rubber itself is a biological product, though human action has modified it greatly by vulcanization. There are elastic tissues in living organisms, the behaviour of which can be interpreted along the lines we have described here. For detailed accounts of theory and experimental studies of rubber and related substances, see Flory (1953, Chapter XI) and Treloar (1958). Both authors give extensive references to the original literature.

CHAPTER 4

Chemical equilibria and Gibbs energy changes of chemical reactions

4.1 Introduction

In any chemical system, including those involved in biological processes, the equilibrium position is of utmost importance. The system will always move spontaneously towards equilibrium, although it will often do so at a significant rate only in the presence of a specific catalyst, e.g. an enzyme.

For example, in the reaction

$$A \rightleftharpoons B$$

the equilibrium position will be defined by the difference in standard Gibbs energy between A and B in their standard states. In turn the usual way to determine the changes in Gibbs energies during chemical transformations is the evaluation of the equilibrium ratio, which is called the equilibrium constant

$$K = [\bar{B}]/[\bar{A}],$$

where $[\bar{A}]$ and $[\bar{B}]$ are called the equilibrium concentrations. It is important to realize that while at equilibrium no net change of reactant concentrations occurs, it is a dynamic state with the reaction proceeding in both directions at the same rate. At this point the Gibbs energy of transformation in either direction is zero.

In the preceding chapters the fundamental laws of thermodynamics have been reviewed and the concept of the chemical potential discussed in detail. In this chapter we shall consider how the laws of thermodynamics and the chemical potential define the equilibrium positions of chemical reactions at constant temperature and pressure and how changes in temperature and pressure perturb such equilibria.

4.2 Chemical equilibrium

At equilibrium (see Figure 4.1) the Gibbs energy of the whole reaction system is at a minimum:

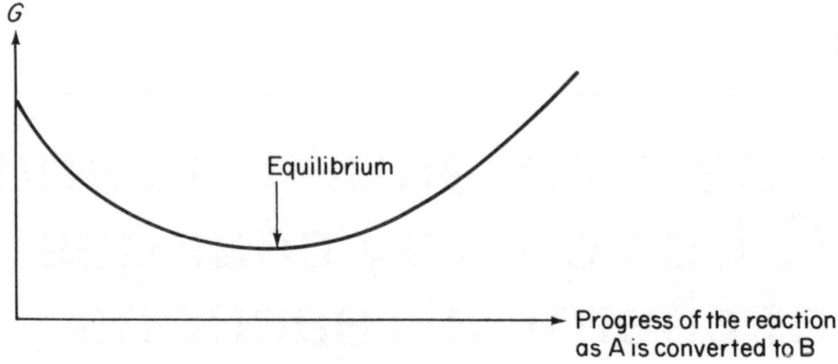

Figure 4.1 Plot of the Gibbs energy of the reaction A $\rightleftharpoons$ B.

$$dx = -d[A] = d[B].$$

As the reaction proceeds from left to right the decrease in Gibbs energy is

$$-dG = -(\mu_B\, dx - \mu_A\, dx)$$

and at equilibrium, when $dG = 0$,

$$\mu_A = \mu_B. \tag{4.1}$$

Applying equation (3.5) one can write

$$\mu_A = \mu_A^\circ + RT \ln [A], \qquad \mu_B = \mu_B^\circ + RT \ln [B],$$

and the standard Gibbs energy difference between the reactants A and B is

$$\Delta G^\circ = \mu_B^\circ - \mu_A^\circ.$$

Therefore at equilibrium, when (4.1) applies,

$$\Delta G^\circ = -RT \ln \frac{[\bar{B}]}{[\bar{A}]} = -RT \ln K. \tag{4.2}$$

Henceforth we shall omit the bar superscripts when denoting concentrations at equilibrium.

Equation (4.1) can be generalized for the case of the reacting system

$$n_A A + n_B B + n_C C + \ldots \rightleftharpoons n_R R + n_S S + n_T T + \ldots,$$

where reactants and products are expressed in capital letters and the respective n_A, etc., are the stoicheiometric numbers. At equilibrium the sum of the chemical potentials of the products times their stoicheiometric numbers is equal to the sum of the chemical potentials of the reactants times their stoicheiometric numbers. This is a simple extension of the principle expressed in equation (4.1). The equilibrium constant for the above system is written in accordance with the law of mass action,

$$K = \frac{[R]^{n_R}[S]^{n_S}[T]^{n_T}\ldots}{[A]^{n_A}[B]^{n_B}[C]^{n_C}\ldots}, \tag{4.3}$$

and is used for the calculation of standard Gibbs energy changes as in equation (4.2).

The theoretical basis for the above statements is fully derived in the discussions of chemical potentials in Chapter 3 and numerical examples will be found throughout the rest of this chapter.

4.3 Some examples of applications of equilibrium measurements to the evaluation of Gibbs energies

The reaction catalysed by the enzyme alcohol dehydrogenase, i.e.

$$NAD^+ + \text{ethanol} \rightleftharpoons NADH + \text{acetaldehyde} + H^+,$$

can be followed by monitoring either the light absorption (at 340 nm) or the fluorescence of NADH. If catalytic amounts of enzyme are added to a solution containing NAD^+ and ethanol, each at 1 mM, as well as buffer to fix the pH at 8, the formation of NADH can be observed until it reaches a steady (equilibrium) concentration. We can write

$$K = \frac{10^{-8}[X]^2}{(10^{-3} - [X])^2},$$

where [X] is the equilibrium concentration of NADH and of acetaldehyde. The measured equilibrium concentration [X] = [NADH] = $[CH_3CHO]$ for the above system at 25 °C, is 2.561×10^{-5} M. This gives

$$K = 6.91 \times 10^{-12} \text{ M}.$$

For a careful investigation of the equilibrium of this reaction the reader is referred to Burton (1974) and references therein. It is important to evaluate the equilibrium position reached when different starting concentrations of the reactants are used and when the equilibrium is approached from both directions.

Burton (1974) also studied the reaction

$$NAD^+ + \text{propan-2-ol} \rightleftharpoons NADH + \text{acetone} + H^+ \tag{4.4}$$

and obtained the equilibrium constant $K = 7.71 \times 10^{-9}$ M (at 25 °C). We can obtain the Gibbs energy as follows:

$$\Delta G^\circ = -RT \ln K = -8.314 \times 298.15 \times \ln(7.71 \times 10^{-9})$$
$$= 46.31 \text{ kJ mol}^{-1}.$$

Since there are good thermochemical data in the literature for the interconversion

$$\text{propan-2-ol (aqueous)} \rightarrow \text{acetone (aqueous)} + H_2 \text{ (gas)}$$

($\Delta G° = 24.4$ kJ mol^{-1}, $\Delta H° = 71.8$ kJ mol^{-1}; see Burton (1974) for references and discussion of range of literature values), it is possible to evaluate the thermodynamic parameters for that ubiquitous biochemical process, the reduction of NAD^+:

$$NAD^+ \text{ (aqueous)} + H_2 \text{ (gas)} \rightleftharpoons NADH \text{ (aqueous)} + H^+ \text{ (aqueous)}$$

($\Delta G° = 22.2$ kJ mol^{-1}, $\Delta H° = -30.6$ kJ mol^{-1}). Burton (1974) evaluated $\Delta H°$ for reaction (4.4) from the slope of the linear regression of ln K against $1/T$ (see Appendix).

Most dehydrogenase reactions linked to the reduction of NAD^+ to NADH have very small equilibrium constants (exceptions will be referred to below). In biological systems one of the products, H^+, is kept fairly constant at low concentration. This results in the other products being formed in measurable amounts.

The reaction

$$\text{glucose 6-phosphate} + NADP^+ \rightleftharpoons \text{gluconolactone 6-phosphate} + NADPH + H^+$$

raises another interesting question. The initial product, gluconolactone 6-phosphate, hydrolyses to gluconate and a hydrogen ion under physiological conditions. Several of the reactions linked to the reduction of $NADP^+$ are coupled to subsequent reactions of the products with water. The consequences of this will be discussed in following sections.

Glaser and Brown (1955) found that at pH 6.4 and 28 °C the reaction of the lactone to form gluconic acid is slow (half time ~ 24 min). At relatively high enzyme concentrations (to obtain rapid equilibration) the equilibrium constant can be determined

$$K = \frac{[\text{gluconolactone 6-phosphate}][NADPH][H^+]}{[\text{glucose 6-phosphate}][NADP^+]} = 6.0 \times 10^{-7}\ \text{M}. \tag{4.5}$$

At neutral pH the reaction proceeds in favour of NADPH and lactone formation.

At pH above neutrality the hydrolysis of the lactone is too fast for the equilibrium to be studied by the methods available to Glaser and Brown. However, rapid reaction techniques can be used to determine equilibria of individual steps as outlined in Section 4.9.

4.4 Dependence of biochemical equilibria on the concentrations of H^+ and other cations

Many biochemical reactions are studied at essentially constant and defined concentrations of certain ions which are involved in the overall equilibrium. These conditions are maintained by the presence of suitable buffers (see Section 5.3).

For reactions of the type catalysed by many dehydrogenases, as for instance that catalysed by alcohol dehydrogenase,

$$NAD^+ + CH_3CH_2OH \rightleftharpoons CH_3CHO + NADH + H^+,$$

the hydrogen ion concentration has to be taken into account as a stoicheiometric reactant. Over the whole pH range the relation for the standard Gibbs energy is given by

$$\Delta G^\circ = -RT \ln \frac{[CH_3CHO][NADH][H^+]}{[CH_3CH_2OH][NAD^+]}$$

$$= -RT\left(\ln \frac{[CH_3CHO][NADH]}{[CH_3CH_2OH][NAD^+]}\right) - RT \ln [H^+] \ .$$

If we define the apparent standard Gibbs energy at a defined pH $= x$ as

$$\Delta G^{\circ\prime}_{(\mathrm{pH}=x)} = -RT \ln \frac{[CH_3CHO][NADH]}{[CH_3CH_2OH][NAD^+]} \ ,$$

then

$$\Delta G^{\circ\prime}_{(\mathrm{pH}=x)} = \Delta G^\circ + RT\, y \ln 10^{-x}, \tag{4.6}$$

where y is the number of protons involved in the stoicheiometry of the reaction. For example, if $T = 298.15$ K, $y = 1$, and pH $= 7$,

$$\Delta G^{\circ\prime}_{(\mathrm{pH}=7)} = \Delta G^\circ - 39.95 \text{ kJ}.$$

Examples of a different type can be introduced with a discussion of the pH dependence of the equilibria

$$R_1COOR_2 + H_2O \overset{K_1}{\rightleftharpoons} R_1COOH + HOR_2 \overset{K_2}{\rightleftharpoons} R_1COO^- + H^+ + HOR_2.$$

The two equilibrium constants are

$$K_1 = \frac{[R_1COOH][HOR_2]}{[R_1COOR_2][H_2O]} \quad \text{and} \quad K_2 = \frac{[R_1COO^-][H^+][HOR_2]}{[R_1COOH][HOR_2]}$$

and the overall equilibrium constant is

$$K = K_1K_2 = \frac{[R_1COO^-][H^+][HOR_2]}{[R_1COOR_2][H_2O]} \, .$$

Since the activity of H_2O is taken as unity in dilute aqueous solutions, this term may be omitted here. If the equilibrium of the reaction is determined at a fixed concentration $[H^+] \simeq K_2$, analysis of the reaction products at equilibrium gives $[R_1COOR_2]$, $[HOR_2]$, and $[R_1COOH] + [R_1COO^-]$. If the apparent equilibrium constant at pH $= x$ is defined as

$$K'_{(pH=x)} = \frac{\{[R_1COO^-] + [R_1COOH]\}[HOR_2]}{[R_1COOR_2]} \, ,$$

It follows that

$$\frac{K'}{K} = \frac{\{[R_1COO^-] + [R_1COOH]\}[HOR_2][R_1COOR_2]}{[R_1COOR_2][R_1COO^-][H^+][HOR_2]}$$

$$= \frac{[R_1COO^-] + [R_1COOH]}{[R_1COO^-][H^+]} \, .$$

Hence the pH-dependent $K'_{(pH=x)}$ obtained from the analysis of all ionic and non-ionic forms of reactants is related to the pH-independent overall equilibrium constant K by

$$K'_{(pH=x)} = K\left(\frac{1}{K_2} + \frac{1}{[H^+]}\right) \tag{4.7}$$

and when $[H^+] \gg K_2$ then $K' = K/K_2 = K_1$ and when $[H^+] \ll K_2$ then $K' = K/[H^+]$.

The apparent standard Gibbs energy for this specified condition is

$$\Delta G^\circ_{(pH=x)} = -RT \ln K'_{(pH=x)}.$$

For the more complex case of peptide bond hydrolysis,

$$\begin{array}{ccccc} R_1CONHR_2 & \overset{K_1}{\rightleftharpoons} & R_1COOH & + & R_2NH_2 \\ & & \updownarrow K_A & & \updownarrow K_B \\ & & R_1COO^- & + & R_2NH_3^+ \end{array}$$

the pH-dependent apparent equilibrium constant is given by

$$K'_{(pH=x)} = K_1\left(1 + \frac{[H^+]}{K_B}\right)\left(1 + \frac{K_A}{[H^+]}\right) .$$

As an example of a related case of the pH dependence of an equilibrium we consider the experimental results of Dobry, Fruton and Sturtevant (1952). They used an isotope dilution method to analyse the concentrations of reactants after enzymic equilibration of the reaction

$$\underset{\text{BT}}{\text{benzoyltyrosine}} + \underset{\text{GA}}{\text{glycinamide}} \rightleftharpoons \underset{\text{BTGA}}{\text{benzoyltyrosyl glycylamide}}.$$

The activity of water is taken as unity and is left out of the equation. Neglecting the hydroxyl group of tyrosine, BTGA is taken as uncharged, BT exists in two forms (BTCOOH and $BTCOO^-$), and GA exists in two forms (H_2NGA and ^+H_3NGA). From the analysis of Dobry *et al.* (1952) at pH 7.9 and 25 °C the equilibrium constant

$$K'_{(\text{pH}=7.9)} = \frac{[\text{BTGA}]}{\{[\text{BTCOOH}] + [\text{BTCOO}^-]\}\{[\text{NH}_2\text{GA}] + [^+\text{H}_3\text{NGA}]\}}$$
$$= 0.256\ \text{M}^{-1} \tag{4.8}$$

was obtained. If the two ionic dissociation constants

$$K_A = [\text{BTCOO}^-]\,[\text{H}^+]/[\text{BTCOOH}] = 10^{-3.7}\ \text{M}$$

and

$$K_B = [\text{NH}_2\text{GA}]\,[\text{H}^+]/[^+\text{H}_3\text{NGA}] = 10^{-7.93}\ \text{M}$$

are used to calculate the concentrations of the ionic forms at pH 7.9, we can evaluate the equilibrium constant:

$$K = [\text{BTGA}]/[\text{BTCOO}^-]\,[^+\text{H}_3\text{NGA}] = 0.49\ \text{M}^{-1}.$$

To calculate K', which is obtained from analysis of all species of the reactants, at different pH, from the pH-independent K, we substitute

$$[\text{BTCOOH}] = [\text{BTCOO}^-]\,[\text{H}^+]/K_A$$

and

$$[\text{H}_2\text{NGA}] = [^+\text{H}_3\text{NGA}]\,K_B/[\text{H}^+]$$

for the non-ionic forms in equation (4.8), and we obtain

$$K'_{(\text{pH}=x)} = \frac{K}{(1 + [\text{H}^+]/K_A)(1 + K_B/[\text{H}^+])}. \tag{4.9}$$

Figure 4.2 shows the pH dependence of K'.

The standard Gibbs energy calculated from

$$\Delta G^\circ = -RT \ln K$$

is 1.76 kJ mol^{-1}. Borsook (1953) reports the very much larger value of 17.3 kJ mol^{-1} for the formation of the peptide bond in alanylglycine from

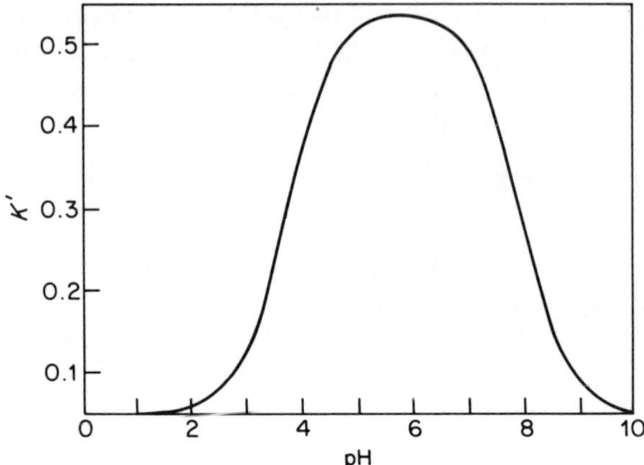

Figure 4.2 The pH dependence of the equilibrium constant K' for the formation of a peptide bond in the reaction given in (4.8). (From Dobry *et al.*, 1952.)

fully ionized reactants to fully ionized products. In the experiments of Dobry *et al.* the Gibbs energy is that of the formation of an uncharged peptide.

A different example is illustrated by the reaction catalysed by fumarase. At neutrality or higher pH, when reactant and product are fully ionized,

$$\underset{\text{fumarate}}{{}^{-}OOC{-}CH{=}CH{-}COO^{-}} \underset{-H_2O}{\overset{+H_2O}{\rightleftharpoons}} \underset{\text{malate}}{CH(OH)COO^{-}{-}CH_2COO^{-}}$$

the equilbrium constant is $K = 4.42$ (at 25 °C).

Krebs (1953) discusses the effects of pH on a number of biochemical equilibria and he derives an equation for the dependence of K' on pH for the fumarase reaction:

$$K' = \frac{T^{M}}{T^{F}} = \frac{A^{M^{2-}}}{A^{F^{2-}}} \frac{K_1^F K_2^F \{K_1^M K_2^M + [H^+]^2 + [H^+]K_1^M\}}{K_1^M K_2^M \{K_1^F K_2^F + [H^+]^2 + [H^+]K_1^F\}}, \tag{4.10}$$

where T^M and T^F are the total concentrations and $A^{M^{2-}}$ and $A^{F^{2-}}$ are the fully ionized concentrations of malate and fumarate respectively. K_1^F, K_2^F, K_1^M, K_2^M are the two ionization constants for fumarate and malate respectively. The ratio $A^{M^{2-}}/A^{F^{2-}}$ can be determined at pH above neutrality and can then be used to calculate K' at any $[H^+]$.

The values for the four dissociation constants used by Krebs were

$$K_1^F = 9.6 \times 10^{-4}\ \text{M}, \qquad K_2^F = 4.0 \times 10^{-5}\ \text{M},$$
$$K_1^M = 3.3 \times 10^{-4}\ \text{M}, \qquad K_2^M = 7.7 \times 10^{-6}\ \text{M}.$$

Substitution of the constants into equation (4.10) results in the values for K' at different values of pH which are illustrated by the graph in Figure 4.3, K'

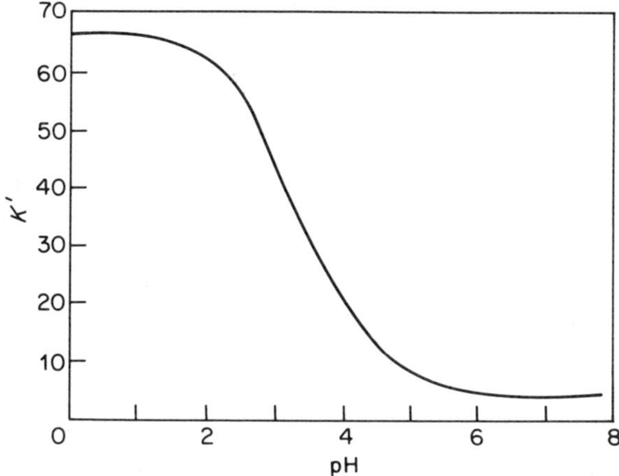

Figure 4.3 The pH dependence of the equilibrium constant K' for the reaction fumarate $\rightleftharpoons$ malate. (From Krebs, 1953.)

having been obtained from an experimentally determined value of the ratio $A^{M^{2-}}/A^{F^{2-}} = 4.40$.

For the equilibrium between two monobasic acids I and II, the equation becomes

$$\frac{T^{I}}{T^{II}} = \frac{A^{I^-}}{A^{II^-}} \frac{K^{II}([H^+] + K^{I})}{K^{I}([H^+] + K^{II})}$$

This equation is useful for the interpretation of equilibria between phosphate esters and between inorganic phosphate and its compounds under conditions when only one of the ionizing groups is titrated – which is always true over the physiological range of pH. The situation is, however, more complex when interactions with other cations also occur (see below) and when phosphate diester bonds are involved in the reaction. The experimental determination of the equilibrium constant for ATP hydrolysis is discussed in Section 4.6. Here we wish to discuss the pH dependence of this process. The reaction written in terms of all ionized forms at specified pH is

$$\text{total ATP} + H_2O \rightleftharpoons \text{total ADP} + \text{total } P_i$$

and thus

$$K'_{(pH=x)} = \frac{[\text{total ADP}][\text{total } P_i]}{[\text{total ATP}]}. \tag{4.11}$$

At high pH (> 8) the predominant reaction becomes

$$ATP^{4-} + H_2O \rightleftharpoons ADP^{3-} + P_i^{2-} + H^+$$

and

$$K = \frac{[ADP^{3-}][P_i^{2-}][H^+]}{[ATP^{4-}]} .$$

This can be interpreted according to equation (4.11). However, at physiological pH the apparent equilibrium constant describing the concentrations of the total (all ionized forms) of reactants is expressed as

$$K'_{(pH=x)} = \frac{\{[ADP^{3-}] + [ADP^{2-}]\}\{[P_i^{2-}] + [P_i^{-}]\}}{[ATP^{4-}] + [ATP^{3-}]} . \qquad (4.11a)$$

Phillips, George and Rutman (1963) determined the pK values for the three reactants of this equilibrium at different temperatures and ionic strengths. At 25 °C and physiological ionic strength (see Figure 4.4 for temperature and ionic strength dependence)

$$K_\alpha = [ATP^{4-}][H^+]/[ADP^{3-}] = 10^{-7.1}\ \text{M},$$
$$K_\beta = [ADP^{3-}][H^+]/[ADP^{2-}] = 10^{-6.8}\ \text{M},$$
$$K_\gamma = [HPO_4^{2-}][H^+]/[H_2PO_4^{-}] = 10^{-6.9}\ \text{M}.$$

Proceeding as before in derivations of the relation between the pH-dependent equilibrium constant K' and K, we substitute

$$[ATP^{3-}] = [ATP^{4-}][H^+]/K_\alpha$$
$$[ADP^{2-}] = [ADP^{3-}][H^+]/K_\beta$$
$$[H_2PO_4^{-}] = [HPO_4^{2-}][H^+]/K_\gamma$$

in equation (4.11a) and obtain

$$K'_{(pH=x)} = K\frac{(1 + [H^+]/K_\beta)(1 + [H^+]/K_\gamma)}{(1 + [H^+]/K_\alpha)[H^+]} . \qquad (4.12)$$

In this equation the hydrogen ion concentration appears in the denominator since a proton is produced in the above reactions. This is similar to the case of ester hydrolysis expressed in equation (4.7).

In addition to different hydrogen ion dissociation equilibria one has to consider the complexes formed with other cations present in biological systems. In connection with the reactions of ATP the intracellular concentration of Mg^{2+} is of particular importance. Many enzymes, contractile proteins, and ion transport ATPases react only with the Mg–ATP complex, although they bind ATP in the absence of Mg. Mg binds strongly to ATP^{4-} and ADP^{3-} but only weakly to orthophosphate. The following data are of interest for comparison, the effects of temperature and ionic strength are given by Rosing and Slater (1972):

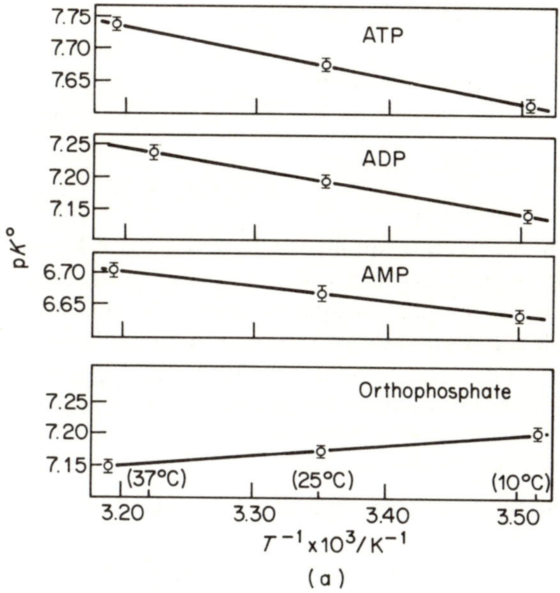

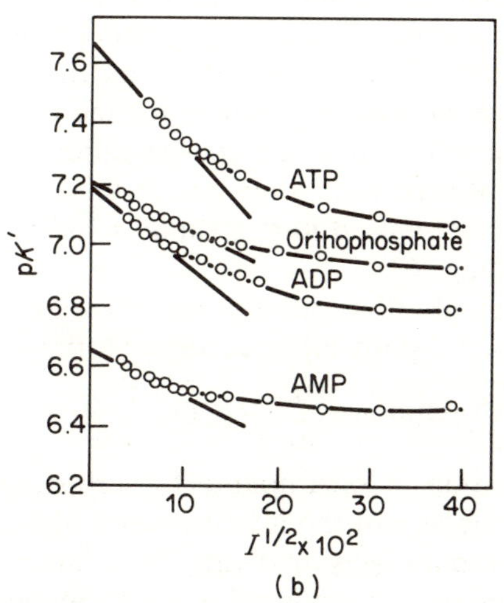

Figure 4.4 (a) p$K°$ for the secondary phosphate ionization of ATP, ADP, AMP, and orthophosphoric acid plotted against T^{-1}. (b) pK' for the secondary phosphate ionizations of ATP, ADP, AMP, and orthophosphoric acid plotted against $I^{1/2}$ at a temperature of 25 °C. Titration was carried out with tetra-*n*-propyl ammonium hydroxide and ionic strength adjusted with the corresponding bromide.The limiting slopes drawn in (b) have the following values: AMP 1.5; ADP 2.5; ATP 3.5; orthophosphate 1.5. (Redrawn from Phillips *et al.*, 1963.)

$$ATP\text{–}Mg^{2-} \rightleftharpoons ATP^{4-} + Mg^{2+}, \quad K = 2.2 \times 10^{-5}\ \text{M at } 37\ ^\circ\text{C}, I = 0.17;$$

$$ADP\text{–}Mg^{-} \rightleftharpoons ADP^{3-} + Mg^{2+}, \quad K = 2.86 \times 10^{-4}\ \text{M at } 37\ ^\circ\text{C}, I = 0.17;$$

$$HPO_4\text{–}Mg \rightleftharpoons HPO_4^{2-} + Mg^{2+}, \quad K = 8.85 \times 10^{-2}\ \text{M at } 37\ ^\circ\text{C}, I = 0.17.$$

The effects on the overall equilibria of reactions of ATP due to the presence of Mg^{2+} can be calculated in a manner analogous to that outlined for the effects of different hydrogen ion concentrations. We shall return to this topic in connection with the discussion of the determination of the Gibbs energy of hydrolysis of ATP. The effects of Mg^{2+} ion concentrations on the equilibrium of this reaction will serve as a numerical example (see Section 4.6).

4.5 Spontaneous reactions of primary products with water

In many enzyme-catalysed equilibria the substrates and products occur in two or more forms, which are in turn in equilibrium with each other. The ionization equilibria discussed in the previous section are examples of such phenomena. Other examples are optical isomer formation, the spontaneous hydrolysis of lactones, and the hydration of primary reaction products. These phenomena have to be taken into account when the data from reaction equilibria are interpreted. They are also of importance for the interpretation of substrate binding equilibria since the substrate concentration is usually reported as the sum of the multiple forms, while only one of these forms may bind or react. Kinetic techniques can be used to determine some of these linked equilibria (see Gutfreund, 1975, p. 63).

An important example of an ubiquitous substrate for biological processes which occurs in several rapidly interconverting forms is CO_2:

$$CO_2(\text{gas}) \rightleftharpoons CO_2(\text{aqueous}) \underset{-H_2O}{\overset{+H_2O}{\rightleftharpoons}} H_2CO_3 \underset{-H^+}{\overset{+H^+}{\rightleftharpoons}} HCO_3^-.$$

The further dissociation of bicarbonate to carbonate at high pH is not of physiological importance. For the interpretation of experimental results from the study of reactions putatively involving CO_2, data concerning the equilibria expressed above need to be considered (see also Section 3.15).

The solubility of CO_2 in water at 1 atm pressure of pure gas is 0.0339 M at 25 °C and this is reduced to 0.0328 M at ionic strength $I = 0.1$. Incidentally this corresponds to a concentration of approximately 10^{-5} M CO_2 dissolved in water in equilibrium with air, which contains about 0.03 vol% of that gas.

The total concentration $\{[CO_2] + [H_2CO_3] + [HCO_3^-]\}$ obtained from analysis at a given pressure of CO_2 gas depends on the pH of the solution. The apparent dissociation constant obtained from pH titration is

$$K'_A = \frac{[HCO_3^-][H^+]}{[CO_2] + [H_2CO_3]} = 4.45 \times 10^{-7} \text{ M}$$

and $pK'_A = 6.352$. The equilibrium constant for the hydration at 25 °C,

$$K_H = [H_2CO_3]/[CO_2] \simeq 0.0026,$$

has been evaluated from kinetic studies with a range of results. However, the true first ionization constant of carbonic acid,

$$K_A = [H^+]\,[HCO_3^-]/[H_2CO_3] = 10^{-3.77} \text{ M},$$

can be determined from high frequency conductivity measurements (Wien effect). For a summary of the results obtained, as well as for detailed references, the review by Edsall (1969) should be consulted. It is then possible to compute a value for K_H from the two dependent equilibria characterized by K'_A and K_A. Inspection of the above definitions for the three constants shows that at 25 °C

$$K'_A/K_A = K_H/(1 + K_H) \simeq K_H = 10^{-2.58}.$$

The approximation is justified since $K_H \ll 1$.

Kinetic techniques were employed to examine whether CO_2, H_2CO_3, or HCO_3^- is the primary substrate or product of such enzymes as urease, yeast carboxylase, 6-phosphogluconate dehydrogenase, and others. Gutfreund (1975, p. 49) summarized the results which demonstrated that CO_2 was the reacting form in the above enzymes. However, carboxylations catalysed by enzymes which have biotin as prosthetic groups utilize HCO_3^- as the primary substrate (see Wimmer and Rose, 1978).

The work of Dalziel and his colleagues should be consulted for the thermodynamic analysis of coupled oxidative decarboxylations (see Landsborough and Dalziel, 1968; Villet and Dalziel, 1969). One of the systems described by Villet and Dalziel will serve here as an example. At constant pH 7 we can write the equilibrium for the reaction catalysed by 6-phosphogluconate dehydrogenase: thence

$$K'_{(pH=7)} = [NADPH]\,[\text{ribulose-5-phosphate}] \times p_{CO_2} / [NAD^+]\,[\text{6-phosphogluconate}] = 2.38 \text{ atm}$$

where p_{CO_2} is the partial pressure of CO_2 in atmospheres at 25 °C. This equilibrium constant does not vary significantly with pH from 6.5 to 7.5. If the partial pressure is converted into concentration of dissolved CO_2, one obtains

$$K'_{(pH=7)} = 2.38 \times 0.03 = 0.072 \text{ M}.$$

Similar investigations are described by Dalziel and his colleagues on the oxidative decarboxylation of malate and isocitrate.

The standard Gibbs energy of the transfer of CO_2 (using 1 atm CO_2 as standard state in the gas phase) from the gaseous to the aqueous phase can be calculated from the solubility as an equilibrium process at 25 °C:

$$K = \frac{[CO_2\ (\text{aqueous})]}{[CO_2\ (\text{gas})]} = 0.034$$

and thus

$$\Delta G^\circ = -RT \ln K = 8.24 \text{ kJ mol}^{-1}.$$

4.6 Group transfer reactions

A large number of chemical reactions in biological systems result in the transfer of acyl or phosphoryl groups from a donor to an acceptor molecule. Such processes are not only of importance in metabolism but also in the control of ion transport, muscle contraction, and the transmission of signals via synapses. The thermodynamics of the transfer of phosphate groups is of particular interest since a large proportion of the energy available due to aerobic (oxidative) and anaerobic (glycolytic) breakdown of nutrients is utilized for the synthesis of a number of key phosphate esters.

Since Lipmann (1941) stressed the importance of ATP in the energy balance of biological systems there have been many arguments about the correct description of 'the utilization of the energy stored' in the γ and β phosphate ester bonds of this compound, which are formed during oxidative phosphorylation or substrate-linked processes (for details see for instance Lehninger, 1975). One problem arises from the frequent use of the term 'high energy' phosphate bond for the two terminal ester linkages of ATP. This has nothing to do with bond energy as understood in chemistry, but refers to the relatively large apparent standard Gibbs energy change of hydrolysis of these bonds. Table 4.1 gives values for the standard Gibbs energy changes ($\Delta G^{\circ\prime}$) of a number of reactions. It can be seen that the value for ATP hydrolysis is approximately midway in the list, among other phosphate esters. This 'buffering' position is significant in systems like muscle where a considerable amount of ATP is hydrolysed in rapid transients and re-phosphorylated by phosphate transfer from phosphocreatine, a process catalysed by the enzyme creatine kinase.

On an historical note it is worthwhile mentioning that Meyerhof and Schulz (1935) found, soon after the discovery of creatine phosphate by Fiske and Subbarow and by Eggleton and Eggleton (see Needham, 1971, for a detailed account), that the heat of hydrolysis of this compound was -46 kJ mol^{-1}. They noted that this value was approximately four times that observed for glucose phosphate ester hydrolysis. This was of great significance in the context of Meyerhof's classical attempts to use muscle contraction as a model for the explanation of physiological energy requirements in terms of the identified

Table 4.1 Standard free energy changes at pH 7: $\Delta G^{\circ\prime}$ at 25 °C for some reactions selected for the discussion of enzyme-catalysed processes

	$\Delta G^{\circ\prime}$/ kcal mol^{-1}	$\Delta G^{\circ\prime}$/ kJ mol^{-1}
Hydrolysis of phosphate esters †		
1,3-Diphosphoglycerate → 3-phosphoglycerate + P_i	−13.6	−56.9
Phosphoenol pyruvate → pyruvate + P_i	−13.3	−55.6
Creatine phosphate → creatine + P_i	−10.2	−42.7
Acetyl phosphate → acetate + P_i	−10.1	−42.3
Adenosine triphosphate → adenosine phosphate + pyrophosphate	− 8.0	−33.5
Adenosine triphosphate → adenosine diphosphate + P_i	− 7.7	−32.2
Adenosine diphosphate → adenosine phosphate + P_i	− 6.6	−27.6
Pyrophosphate → 2 P_i	− 6.6	−27.6
Glucose 1-phosphate → glucose + P_i	− 5.0	−20.9
Glucose 6-phosphate → glucose + P_i	− 3.3	−13.8
Fructose 6-phosphate → fructose + P_i	− 3.1	−12.9
Glycerol 1-phosphate → glycerol + P_i	− 2.3	− 9.6
Some other reactions		
Acetyl coenzyme A → acetate + coenzyme A	− 8.0	−33.5
NAD^+ + H_2(gas) → NADH + H^+	+ 5.31	+22.2
Ethanol → acetaldehyde	+ 9.68	+40.5

†P_i stands for orthophosphate.

chemical reactions in this system (Meyerhof, 1930). It must be emphasized that these were measurements of ΔH and not of ΔG!

One can consider the hydrolysis of an acyl or phosphate ester as a special case of transfer to water as the acceptor. The standard Gibbs energy change of hydrolysis is larger than that obtained from transfer to other acceptors and is called the transfer potential. The term 'high transfer potential' is therefore to be preferred to 'high energy bond' for the phosphoryl bond of ATP and the compounds above ATP in Table 4.1. Clearly the energy change of hydrolysis is the sum of the energy changes of many processes involving bond breaking, bond formation, ionization, and solvation. Lipmann (1941) himself has, of course, appreciated that some term like transfer potential is more suitable than bond energy for the relatively high Gibbs energy of hydrolysis of some phosphate esters.

The direct determination of the equilibrium constant for the reaction

$$\text{ATP} + \text{H}_2\text{O} \overset{K}{\rightleftharpoons} \text{ADP} + \text{P}_\text{i}$$

can not be performed with reasonable accuracy. In this case there are no suitable conditions under which the enzyme-catalysed reaction would, on equilibration, produce significant concentrations of all the reactants. At equilibrium the reaction lies very far to the right. The standard Gibbs energy of this reaction, which is of key importance in the evaluation of the energy balance of many biological processes, has therefore been determined in coupled systems. A number of authors have studied the reaction system catalysed by the two enzymes glutamine synthetase and glutaminase:

$$\text{glutamate} + \text{ammonia} + \text{ATP} \overset{K_1}{\rightleftharpoons} \text{glutamine} + \text{ADP} + \text{P}_\text{i}$$

and

$$\text{glutamine} + \text{H}_2\text{O} \overset{K_2}{\rightleftharpoons} \text{glutamate} + \text{ammonia}.$$

The product of the two equilibria gives

$$K_1K_2 = [\text{ADP}]\,[\text{P}_\text{i}]/[\text{ATP}] = K'.$$

Rosing and Slater (1972) undertook a critical examination of previous reports, as well as their own studies of the above reactions, and took into account the best available dissociation constants for H^+ and Mg^{2+} for all the reactants in the above equilibria. In this way they obtained values for the apparent standard Gibbs energy of the formation of ATP from ADP and P_i under physiological conditions and in solutions of varied compositions.

It was pointed out above that the apparent equilibrium constants and standard Gibbs energies for such reactions depend upon the ionic composition of the solutions. The theory and relevant examples have been discussed in Section 4.4. Here we present the data obtained by Rosing and Slater (1972) for the H^+ and Mg^{2+} dependence of $\Delta G^{\circ\prime}$, the apparent standard Gibbs energy, in Table 4.2 and Figure 4.5.

Table 4.2 Values of $-\Delta G°$ in kilocalories per mole at different values of pH, ionic strength, and Mg^{2+} ion concentration for the hydrolysis of ATP at 25 °C and 37 °C. The values in parentheses are in kilojoules per mole. (From Rosing and Slater, 1972)

pH	I	$[Mg^{2+}]$/mM				
		0	1.0	10.0	25.0	50.0
				25 °C		
6.00	0.00	7.59 (31.77)				
6.00	0.10	7.25 (30.33)	6.58 (27.53)	6.17 (25.82)	6.18 (25.85)	
6.00	0.15	7.30 (30.54)	6.64 (27.76)	6.23 (26.05)	6.23 (26.05)	6.30 (26.34)
6.00	0.20	7.37 (30.86)	6.69 (27.97)	6.29 (26.32)	6.30 (26.34)	6.38 (26.08)
6.50	0.00	7.71 (32.27)				
6.50	0.10	7.44 (31.12)	6.51 (27.25)	6.31 (26.39)	6.42 (26.87)	
6.50	0.15	7.49 (31.33)	6.57 (27.49)	6.36 (26.60)	6.46 (27.02)	6.62 (27.69)
6.50	0.20	7.56 (31.62)	6.62 (27.70)	6.43 (26.91)	6.54 (27.38)	6.71 (28.09)
7.00	0.00	8.01 (33.51)				
7.00	0.10	7.82 (32.72)	6.69 (28.00)	6.66 (27.88)	6.87 (28.74)	
7.00	0.15	7.87 (32.91)	6.75 (28.25)	6.71 (28.06)	6.90 (28.85)	7.13 (29.82)
7.00	0.20	7.93 (33.17)	6.80 (28.46)	6.79 (28.40)	6.99 (29.25)	7.23 (30.26)
7.50	0.00	8.55 (35.77)				
7.50	0.10	8.38 (35.05)	7.12 (29.80)	7.20 (30.12)	7.46 (31.20)	
7.50	0.15	8.42 (35.23)	7.18 (30.06)	7.24 (30.28)	7.48 (31.28)	7.74 (32.40)
7.50	0.20	8.48 (35.48)	7.23 (30.27)	7.32 (30.04)	7.58 (31.70)	7.85 (32.85)
8.00	0.00	9.24 (38.67)				
8.00	0.10	9.02 (37.74)	7.71 (32.24)	7.83 (32.74)	8.11 (33.91)	
8.00	0.15	9.06 (37.90)	7.77 (32.51)	7.86 (32.90)	8.12 (33.99)	8.40 (35.16)
8.00	0.20	9.12 (38.16)	7.82 (32.71)	7.95 (33.26)	8.22 (34.41)	8.51 (35.62)
8.50	0.00	9.96 (41.67)				
8.50	0.10	9.69 (40.54)	8.35 (34.96)	8.49 (35.52)	8.78 (36.72)	
8.50	0.15	9.73 (40.70)	8.42 (35.22)	8.53 (35.68)	8.80 (36.82)	9.09 (38.03)
8.50	0.20	9.79 (40.96)	8.47 (35.42)	8.61 (36.04)	8.90 (37.22)	9.19 (38.45)
9.00	0.00	10.66 (44.59)				
9.00	0.10	10.37 (43.38)	9.03 (37.76)	9.17 (38.36)	9.46 (39.56)	
9.00	0.15	10.41 (43.54)	9.09 (38.03)	9.20 (38.50)	9.47 (39.63)	9.76 (40.85)
9.00	0.20	10.47 (43.80)	9.14 (38.23)	9.29 (38.87)	9.58 (40.06)	9.87 (41.31)
				37 °C		
6.00	0.00	7.69 (32.16)				
6.00	0.10	7.25 (30.34)	6.48 (27.10)	6.14 (25.70)	6.21 (25.98)	
6.00	0.15	7.27 (30.41)	6.48 (27.10)	6.14 (25.70)	6.21 (25.98)	6.34 (26.51)
6.00	0.20	7.32 (30.65)	6.48 (27.10)	6.19 (25.88)	6.27 (26.22)	6.42 (26.85)
6.50	0.00	7.81 (32.69)				
6.50	0.10	7.48 (31.30)	6.46 (27.02)	6.36 (26.63)	6.56 (27.44)	
6.50	0.15	7.51 (31.41)	6.47 (27.07)	6.37 (26.65)	6.55 (27.40)	6.78 (28.35)
6.50	0.20	7.57 (31.67)	6.50 (27.18)	6.44 (26.95)	6.64 (27.80)	6.89 (28.82)
7.00	0.00	8.13 (34.00)				
7.00	0.10	7.92 (33.15)	6.71 (28.09)	6.81 (28.51)	7.10 (29.70)	
7.00	0.15	7.95 (33.28)	6.74 (28.22)	6.82 (28.54)	7.09 (29.65)	7.37 (30.85)
7.00	0.20	8.03 (33.59)	6.79 (28.43)	6.92 (28.94)	7.20 (30.13)	7.50 (31.38)
7.50	0.00	8.70 (36.40)				
7.50	0.10	8.53 (35.71)	7.21 (30.18)	7.42 (31.03)	7.75 (32.41)	
7.50	0.15	8.57 (35.85)	7.26 (30.36)	7.43 (31.07)	7.73 (32.34)	8.04 (33.65)
7.50	0.20	8.65 (36.18)	7.32 (30.63)	7.53 (31.52)	7.86 (32.87)	8.18 (34.21)
8.00	0.00	9.44 (39.49)				
8.00	0.10	9.22 (38.56)	7.84 (32.81)	8.09 (33.85)	8.43 (35.29)	
8.00	0.15	9.25 (38.70)	7.89 (33.02)	8.10 (33.88)	8.42 (35.23)	8.74 (36.56)
8.00	0.20	9.33 (39.04)	7.96 (33.32)	8.21 (34.35)	8.55 (35.76)	8.88 (37.14)

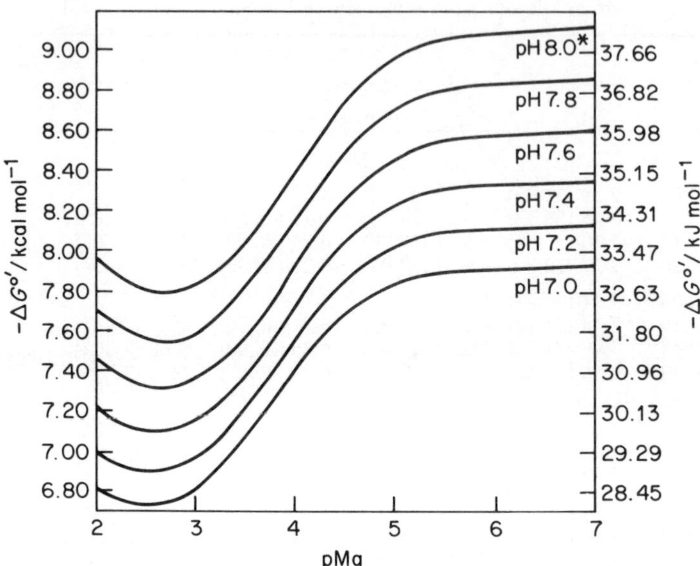

Figure 4.5 The pH and Mg^{2+} ion concentration dependence of $\Delta G^{o\prime}$ for the hydrolysis of ATP at 25 °C and $I = 0.2$. (From Rosing and Slater, 1972.)

The results of Rosing and Slater are in reasonable agreement with values obtained for the equilibrium constant of the hydrolysis of ATP obtained by Robbins and Boyer (1957) from another coupled system:

$$\text{glucose} + \text{ATP} \xrightleftharpoons{\text{hexokinase}} \text{glucose 6-phosphate} + \text{ADP}$$

and

$$\text{glucose 6-phosphate} \xrightleftharpoons{\text{phosphatase}} \text{glucose} + P_i.$$

An interesting role of enzymes during the transfer of phosphoryl or acyl groups is the preservation of the transfer potential in the enzyme–substrate complex. Two possible schemes are available for the transfer of a phosphate or an acyl group from a donor to an acceptor, avoiding the loss of Gibbs energy accompanied by hydrolysis: (1) direct transfer of the group with both donor and acceptor bound to the enzyme; (2) the substituted enzyme mechanism in which the group is transferred from donor to protein and from protein to acceptor. Details of some of the chemical events and proposed structures for enzyme intermediates are found in treatments of enzyme mechanisms (see for instance Fersht, 1977). In this connection the treatment of the relation between substrate binding and group transfer energy given by Jencks (1975) is also very relevant.

Clearly reactions like those catalysed by hexokinase have to be coupled via an enzyme intermediate. Although the enzyme does not affect the overall equilibrium

$$\text{ATP} + \text{glucose} \rightleftharpoons \text{ADP} + \text{glucose 6-phosphate},$$

for which

$$\Delta G^{\circ\prime}_{(\text{pH}=7)} \simeq (13.8 - 32.9) = -19.1 \text{ kJ mol}^{-1}$$

and

$$K'_{(\text{pH}=7)} \simeq e^{-\Delta G^{\circ\prime}/RT} \simeq 2.23 \times 10^3,$$

this reaction will proceed only in the presence of Mg^{2+} and should strictly be written in terms of nucleotides complexed with the divalent ion. If one were to perform an experiment with two enzymes catalysing the equilibration of the two separate reactions

$$\text{ATP} + H_2O \overset{\text{ATPase}}{\rightleftharpoons} \text{ADP} + P_i$$

and

$$\text{glucose 6-phosphate} \overset{\text{glucose 6-phosphatase}}{\rightleftharpoons} \text{glucose} + P_i$$

the concentrations of the end products glucose 6-phosphate and glucose would be quite different from those resulting from the kinase reaction starting with the same initial concentration of glucose and ATP. Starting with 10^{-3} M glucose and 10^{-3} M ATP, the hexokinase reaction would yield at equilibrium a concentration x of glucose 6-phosphate:

$$[\text{ADP}]\,[\text{glucose 6-phosphate}]/[\text{ATP}]\,[\text{glucose}] = 2.23 \times 10^3;$$

therefore

$$x^2/(10^{-3} - x)^2 = 2.23 \times 10^3$$

and

$$x = 0.98 \times 10^{-3} \text{ M}.$$

On the other hand equilibration of glucose and ATP with ATPase and glucose 6-phosphatase would yield less than 10^{-9} M glucose 6-phosphate. In this discussion we neglect any artefacts which can occur due to side reactions, such as the hydrolysis of ATP by kinases and the direct transfer of phosphate by hydrolyses to acceptors other than water.

The phenomenon of transfer via enzymes or coenzymes to preserve the potential, illustrated above for phosphate, is of considerable importance in a wide range of metabolic and synthetic pathways. Another example is the transfer of acetyl groups (a thiol ester) via an SH group on coenzyme A, for example in the reactions catalysed by citrate synthetase:

$$\text{acetyl CoA} + \text{oxaloacetate} \xrightarrow{\text{citrate synthetase}} \text{citrate}.$$

4.7 Concentration gradients, sequential reactions, and Gibbs energies

The values for standard or apparent standard Gibbs energy changes are required as basic constants for the evaluation of the Gibbs energy changes under actual conditions appertaining to physiological systems. In a cell the reactants are not at equilibrium concentrations although they are often not far removed from them. It can be said that a cell at equilibrium is dead, while in an active cell steady state processes are the rule. The actual Gibbs energies of reactions *in vivo* depend on the local concentration ratios. For instance, the standard Gibbs energy change of transferring 1 mol of phosphate from ATP to creatine to form ADP and creatine phosphate, or its reverse, is defined when all four reaction partners are maintained at molar concentration (strictly unit activity): $\Delta G^{\circ}_{(\mathrm{pH}=7)} = 10.5\ \mathrm{kJ\ mol^{-1}}$ at 25 °C. However, at equilibrium, when

$$[\mathrm{ADP}]\,[\text{creatine phosphate}]/[\mathrm{ATP}]\,[\text{creatine}] = e^{-10.5/2.478} = 0.0145,$$

the Gibbs energy of interconversion is zero. At a particular concentration ratio Q, where

$$Q = \frac{[\mathrm{ADP}][\text{creatine phosphate}]}{[\mathrm{ATP}][\text{creatine}]},$$

the Gibbs energy of interconversion is given by

$$\Delta G' = \Delta G^{\circ} + RT \ln Q. \tag{4.12}$$

This equation reduces to $\Delta G' = 0$ if Q is equal to the equilibrium ratio, that is $Q = K$, since $\Delta G^{\circ} = -RT \ln K$ (see p. 124). In general for a reaction such as

$$a\mathrm{A} + b\mathrm{B} + \ldots \rightleftharpoons m\mathrm{M} + n\mathrm{N} + \ldots,$$

with reactants and products at specified concentrations, we can write $\Delta G'$ as

$$\begin{aligned}\Delta G' &= \Delta G^{\circ} + RT \ln ([\mathrm{M}]^m[\mathrm{N}]^n \ldots/[\mathrm{A}]^a[\mathrm{B}]^b \ldots)\\ &= -RT \ln K + RT \ln ([\mathrm{M}]^m[\mathrm{N}]^n \ldots/[\mathrm{A}]^a[\mathrm{B}]^b \ldots).\end{aligned} \tag{4.13}$$

If the second term on the right is equal to $RT \ln K$, $\Delta G' = 0$ and the system is obviously at equilibrium. If, however, this term is negative and is numerically larger than $RT \ln K$, then $\Delta G'$ is negative and the process is at least potentially spontaneous; that is, it will go in the presence of a suitable catalyst. Moreover $\Delta G'$ in this case is a measure of the maximum useful work that the process can supply.

In this calculation we are of course assuming that the concentrations (or chemical potentials) of reactants and products are being maintained constant as the reaction proceeds. This means that fresh reactants must be continually fed into the system, while products are being constantly removed, so as to keep the concentrations unchanged. In other words we are considering a system, not at equilibrium, but in a steady state of turnover. Such states are of course characteristic of living organisms, so that this calculation of Gibbs energy changes is meaningful and useful for biochemical processes, although it takes us beyond the realm of equilibrium thermodynamics.

An interesting example is obtained from calculations of the Gibbs energy of the formation of ATP from ADP and phosphate at concentrations estimated in muscle, which are [ATP] = 10^{-3} M, [ADP] = 10^{-5} M, [phosphate] = 10^{-3} M. Taking the value of $\Delta G^{\circ}_{(\mathrm{pH}=7.2)} = 30$ kJ mol^{-1}, in a medium of physiological ionic composition, we obtain

$$\Delta G' = 30 + RT \ln \frac{10^{-3}}{10^{-5} \times 10^{-3}} = 30 + 2.58 \times 11.5 = 60 \text{ kJ mol}^{-1}$$

at 37 °C (310 K) for the synthesis of ATP.

Conversely approximately 60 kJ mol^{-1} of Gibbs energy become available during the hydrolysis of 1 mol of ATP under physiological conditions. The uncertainties surrounding this value are connected with the difficulties of the determination of the precise composition of living cells.

The equilibrium of the reaction with one substrate and two products becomes much more favourable for product as the total concentration is lowered. This is illustrated for the reaction catalyzed by aldolase in Table 4.3.

If one continues along the sequence of glycolysis one finds that the apparently energetically unfavourable reaction catalysed by glyceraldehyde 3-phosphate dehydrogenase to form 1,3-diphosphoglycerate,

$$\text{glyceraldehyde 3-phosphate} + \text{phosphate} + \text{NAD}^+ \rightleftharpoons \text{1,3-diphosphoglycerate} + \text{NADH},$$

for which

$$K'_{(\mathrm{pH}=7)} = \mathrm{e}^{-\Delta G/RT} = \mathrm{e}^{-6.28/RT} = 0.079 \text{ M}^{-1},$$

is coupled to the removal of products by energetically favourable reactions.

Table 4.3 Effect of total concentration of reactants on the proportion of triose at equilibrium

Initial concentration of fructose 1,6-bisphosphate/M	Equilibrium concentrations/M		
	Fructose 1,6-bisphosphate	Glyceraldehyde 3-phosphate	Dihydroxyacetone phosphates
1.0	0.989	1.08×10^{-2}	1.08×10^{-2}
0.001	0.711×10^{-3}	0.289×10^{-3}	0.289×10^{-3}
0.0001	0.357×10^{-4}	0.322×10^{-4}	0.322×10^{-4}
0.1†	0.092	0.007	0.015
0.001†	0.45×10^{-3}	0.048×10^{-3}	1.05×10^{-3}
0.0001†	0.0113×10^{-3}	0.0073×10^{-3}	0.17×10^{-3}

†Triosephosphate isomerase as well as aldolase added for equilibration.

Similar interpretations apply to sequential reactions during ligand binding and consequential conformation changes (see Section 5.6). It must also be pointed out that the concentrations of enzymes and other functional proteins in biological systems are often so high that reactant protein complexes have to be considered when writing stoicheiometric equations. This latter point is discussed in some detail later in this chapter (see Section 4.9).

4.8 Reaction equilibria, reaction path, and enzyme catalysis

The equilibrium constant for the reaction

$$\mathrm{A} \rightleftharpoons \mathrm{B}$$

can be expressed in terms of the equilibrium concentrations

$$K = [\mathrm{B}]/[\mathrm{A}]$$

or in terms of the kinetics of the balanced reactions

$$\mathrm{d}[\mathrm{B}]/\mathrm{d}t = k_1[\mathrm{A}] - k_{-1}[\mathrm{B}],$$
$$\mathrm{d}[\mathrm{A}]/\mathrm{d}t = k_{-1}[\mathrm{B}] - k_1[\mathrm{A}].$$

At equilibrium,

$$\mathrm{d}[\mathrm{B}]/\mathrm{d}t = \mathrm{d}[\mathrm{A}]/\mathrm{d}t = 0 \quad \text{and} \quad [\mathrm{B}]/[\mathrm{A}] = k_1/k_{-1} = K.$$

In a two-step process

$$\mathrm{A} \overset{K_1}{\rightleftharpoons} \mathrm{B} \overset{K_2}{\rightleftharpoons} \mathrm{C}$$

the equilibrium constant for the overall reaction is the product

$$K_1K_2 = [\mathrm{B}][\mathrm{C}]/[\mathrm{A}][\mathrm{B}] = [\mathrm{C}]/[\mathrm{A}].$$

Since the total Gibbs energy of $\mathrm{A} \rightleftharpoons \mathrm{C}$ must be the sum of the Gibbs energies for the two individual steps,

$$\Delta G^\circ = \Delta G_1^\circ + \Delta G_2^\circ = -RT \ln (K_1K_2)$$

and therefore

$$K = K_1K_2$$

For a cyclic process such as

$$\begin{array}{ccc} \mathrm{A} & \overset{K_1}{\rightleftharpoons} & \mathrm{B} \\ K_4 \updownarrow & & \updownarrow K_2 \\ \mathrm{D} & \underset{K_3}{\rightleftharpoons} & \mathrm{C} \end{array}$$

it can be readily seen that

$$K = [\mathrm{C}]/[\mathrm{A}] = K_1K_2 = K_3K_4.$$

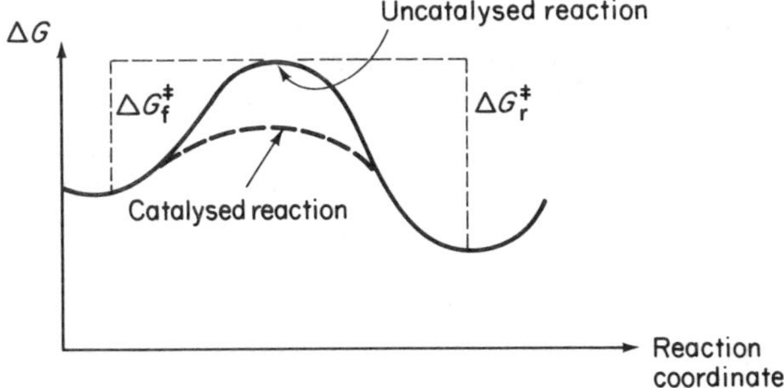

Figure 4.6 The effect of a catalyst on the Gibbs energy of activation of a reaction

This is another manifestation of the characteristics of a state function (see Section 2.3). Other examples of state functions are internal energy, volume, pressure, density, etc. Other properties are called 'path functions', because quantities like heat or work done on a system or by a system are related to the path by which the changes proceed.

There are a number of practical situations where kinetic measurements rather than equilibrium measurements are necessary for the determination of thermodynamic parameters. Typical examples are the kinetic determination of 'on-enzyme equilibria' of individual steps (see Section 4.9) and the method developed by Winkler-Oswatitsch and Eigen (1979) to determine equilibrium constants and enthalpy changes of elementary steps by temperature jump relaxation techniques.

A catalyst is defined as a compound which is present at concentrations which are low compared with the reactants, does not need to be considered in the stoicheiometric equations, and affects the rate constants for the forward and reverse reaction by the same factor. The catalyst changes the Gibbs energy of activation but not the Gibbs energy of the reaction (Figure 4.6).

The difference between the standard Gibbs energy of activation for the forward and reverse reaction is expressed as

$$\Delta G_f^{o\ddagger} - \Delta G_r^{o\ddagger} = \Delta G^o.$$

For a detailed discussion of the activation parameters ($\Delta G^{o\ddagger}$, $\Delta H^{o\ddagger}$, and $\Delta S^{o\ddagger}$) the reader is referred to text-books of physical chemistry (e.g. Moore, 1963; Atkins, 1978).

In general enzymes are correctly considered as catalysts. It must, however, be borne in mind that the above definitions only apply when the enzyme concentration is low (catalytic concentration) compared to the reactants. The consequences of high (stoicheiometric) enzyme concentrations will be considered in the next section.

One important result of the law of enzyme catalysis, as defined above, is that an effector (inhibitor or activator) of an enzyme will influence the forward and reverse rate by the same factor – the ratio has to remain the same to keep the equilibrium constant unchanged. For a linear sequence of reversible reactions effectors can provide little control over the balance of the reaction.

Perusal of text-books of biochemistry published during the last 25 years shows that for one after another of the main biosynthetic pathways the old idea that biosynthesis was a reversal of degradation was found to be erroneous. Fatty acid synthesis does not occur through a reversal of fatty acid oxidation; glycogen synthesis does not occur through a reversal of the phosphorylase reaction; protein synthesis does not occur by a reversal of hydrolysis; and so on. In some cases, as in glycolysis and gluconeogenesis, it is not the entire pathway which is different, but cyclic processes occur at some stage, as for instance

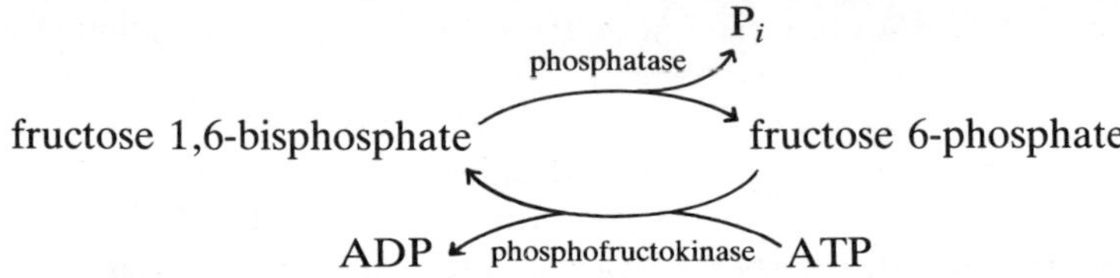

It must be emphasized that, of course, the existence of cycles does not affect the equilibrium. Extension of the argument developed at the beginning of this section shows that the standard Gibbs energy change and hence the equilibrium constant is independent of pathway. The additional flexibility provided by a cyclic system can be illustrated by the control mechanism operating in glycolysis through the above 'minicycle'. Adenosine monophosphate (AMP) activates the kinase and inhibits the phosphatase. When the ATP concentration is low, AMP is high and the rate of glycolysis is increased without loss of the bisphosphate through hydrolysis. High ATP concentration inhibits the kinase but not the phosphatase, thus increasing the rate of gluconeogenesis. Other metabolites also affect this system and an even more sophisticated control system operates in glycogen synthesis and utilization (see Newsholme and Start, 1973). The control of such cycles costs ATP and thus dissipates energy and results in heat production.

The importance of enzymes in biological systems is that they make it possible to establish equilibria very fast, while in their absence most metabolic reactions would proceed at a negligible rate. Also enzyme-catalysed reactions are highly specific; there is almost no waste of materials through side reactions, as in most of the reactions in an organic chemical laboratory. While an extensive treatment of the kinetic behaviour of enzyme reactions is beyond our present scope (see Gutfreund, 1975), some of its features are of importance in the study of biochemical equilibria. The present discussion is initially simplified by consideration of enzymes with one substrate and one product and with each active site independent of all others. Although the equations have to be extended for more complex systems, the principles remain the same. Whatever the number of intermediates in the process

$$\mathrm{E} + \mathrm{S} \underset{k_{-1}}{\overset{k_1}{\rightleftharpoons}} \mathrm{ES} \rightleftharpoons \ldots \rightleftharpoons \mathrm{EP} \underset{k_{-n}}{\overset{k_n}{\rightleftharpoons}} \mathrm{E} + \mathrm{P}, \tag{4.14}$$

$$\begin{array}{c} \mathrm{HC{-}COO^-} \\ \| \\ \mathrm{^-OOC{-}CH} \\ \text{fumarate} \end{array} \quad \underset{-\mathrm{H_2O}}{\overset{+\mathrm{H_2O}}{\rightleftharpoons}} \quad \begin{array}{c} \mathrm{CH(OH)COO^-} \\ | \\ \mathrm{CH_2\ COO^-} \\ \text{malate} \end{array}$$

when the substrate concentration is large compared to enzyme and the product concentration is essentially zero, a steady state rate can be established which is described by an equation of the form

$$v = \frac{d[P]}{dt} = \frac{V[S][E]}{K_m + [S]}, \tag{4.15}$$

where V is the maximum velocity at saturating [S] and K_m is the Michaelis constant. The relation between v and [S] is described by a rectangular hyperbola (see Section 5.1). The two constants K_m and V (defining the reaction of an enzyme with a particular substrate in a solution of defined ionic composition) are functions of a number of rate constants. The more complex the mechanism the more complex are these functions (see for instance Cornish-Bowden, 1976, 1980; Gutfreund 1975, p. 132).

The overall equilibrium constant of the reaction is the product of all the intermediate equilibrium constants:

$$K = \prod_{i=1}^{n} K_i. \tag{4.16}$$

Haldane (1930) showed that the overall equilibrium can be expressed in terms of V and K_m for the forward and reverse reactions:

$$K = V^f K_m^r / V^r K_m^f. \tag{4.17}$$

An interesting consequence of equation (4.17) is that the ratio of the maximum velocities is related to the equilibrium constant by the equation

$$V^f/V^r = K\, K_m^f/K_m^r.$$

For a derivation of the Haldane equation (4.17) see Gutfreund (1975, p. 136).

4.9 The consequences of high enzyme concentrations

In a significant number of organized systems enzymes or related functional proteins (for instance myosin and transport ATPases) occur locally at concentrations comparable with those of their substrates. The concentration of binding sites for some of the intermediates of the glycolytic pathway is such that the release of protein-bound intermediates can be rate limiting for their utilization. One important practical consequence of such situations is the care which has to be taken in the analysis of data for the evaluation of Gibbs energy gradients in metabolic sequences. The concentration of enzymes and of the different complexes can no longer be regarded as negligible (catalytic concentrations); instead they should be regarded as stoicheiometric reactants. It is, therefore, necessary to take into account not only the concentrations of free metabolites but also those of their complexes, when calculations of equilibrium constants or Gibbs energies are performed (see for instance the investigation described by Veech, Raijam, Dalziel, and Krebs, 1969).

The reader is also referred to the later studies by Veech, Lawson, Cornell, and Krebs (1979), which emphasized the high activities of some enzymes in several tissues.

Two types of equilibria concerning enzyme–substrate complexes are of interest; as will be seen they are, of course, coupled. First there are the non-covalent complexes of substrate and product binding and secondly there are the equilibria of the chemical reactions occurring between the enzyme complexes. For the latter one can use the term 'on-enzyme equilibrium' (Gutfreund and Trentham, 1975), and this can be well illustrated by some examples.

Two methods, each suitable for different circumstances, have been used to determine 'on-enzyme equilibria' directly. If the binding equilibria and the overall equilibrium are known one can infer the equilibrium of the chemical step indirectly. One approach to this problem can be applied when the enzyme–substrate and enzyme–product complexes occur at detectable concentrations when the whole system is in true equilibrium. For instance, when the reaction

$$\text{MgATP} + \text{arginine} \xrightleftharpoons{\text{arginine kinase}} \text{MgADP} + \text{arginine phosphate} \tag{4.18}$$

is carried out at a catalytic concentration of arginine kinase at pH 7.25 and 12 °C, the equilibrium constant $K = 0.1$ is obtained. If the same reaction is studied at high enzyme concentrations, ^{31}P nuclear magnetic resonance can be used to follow the rate of phosphate transfer on the enzyme surface. The transfer rate from ATP to arginine is 192 s^{-1} and that from arginine phosphate to ADP is 154 s^{-1}, resulting in an 'on-enzyme' equilibrium with $K^B = 1.25$ for the reaction given in equation (4.18). This approach has been pioneered in Mildred Cohn's laboratory (Rao, Buttlaire and Cohn, 1976; Rao, Cohn and Noda, 1978) and some of her results are given in Table 4.4.

In some enzyme systems the equilibrium between enzyme–substrate and enzyme–product complexes can be observed, although it only occurs transiently. Gutfreund and Trentham (1975) have summarized how transient

Table 4.4 Overall equilibrium constants (K) and interconversion step equilibrium constants (K^B)

Kinase enzymes	$K =$	$K^B =$
Arginine	0.1	1.5
Creatine	0.1	~1
Adenylate	0.4	1.5
Pyruvate (S = pyruvate)	$\sim 3 \times 10^{-4}$	1.0–2.0
Pyruvate (S = glycolate)	$\geqslant 50$	1.0–3.0
3-Phosphoglycerate	$\sim 3 \times 10^{-4}$	1.0

kinetic techniques can be used for this purpose when the dissociation of products is slower than the chemical equilibration on the enzyme surface. Two examples of the application of this approach are the reactions catalysed by heart lactate dehydrogenase and by myosin ATPase. In the following sequence of events,

$$\begin{array}{l} \mathrm{E + NAD^+ + lactate} \\ \quad\quad \updownarrow K_1 \\ \mathrm{E \cdot NAD^+ \cdot lactate} \overset{K^{\mathrm{B}}}{\rightleftharpoons} \mathrm{E \cdot NADH \cdot pyruvate \cdot H^+} \\ \quad\quad\quad\quad\quad\quad\quad\quad\quad\quad \updownarrow K_2 \\ \quad\quad\quad\quad\quad\quad\quad\quad\quad \mathrm{E + NADH + pyruvate + H^+} \end{array}$$

catalysed by lactate dehydrogenase, the rate of dissociation of the ionized complex to the products on the bottom right is slower than the equilibration of the step characterized by K^{B}. When enzyme, $\mathrm{NAD^+}$, and lactate are mixed, all at high concentration, in a stopped-flow apparatus, the transient equilibration of the chemical interconversion

$$\mathrm{E \cdot NAD^+ \cdot lactate} \rightleftharpoons \mathrm{E \cdot NADH \cdot pyruvate \cdot H^+}$$

can be observed by transmission and fluorescence spectroscopy. These experiments are described in detail by Whitaker, Yates, Bennett, Holbrook, and Gutfreund (1974). A comparison of this equilibrium with the equilibrium of the overall reaction at catalytic enzyme concentration gives the following results at a fixed concentration of $[\mathrm{H^+}] = 10^{-7}$ M:

$$K^{\mathrm{B}} = \frac{[\mathrm{E{\cdot}NADH{\cdot}pyruvate{\cdot}H^+}]}{[\mathrm{E{\cdot}NAD^+{\cdot}lactate}]} = 0.25, \tag{4.19}$$

$$K'_{(\mathrm{pH}=7)} = \frac{[\mathrm{pyruvate}][\mathrm{NADH}]}{[\mathrm{lactate}][\mathrm{NAD^+}]} = 0.37 \times 10^{-4}. \tag{4.20}$$

Similarly, for the hydrolysis of ATP by myosin (at pH 8 and $[\mathrm{Mg^{2+}}] = 5$ mM),

$$K^{\mathrm{B}} = \frac{[\mathrm{M{\cdot}ADP{\cdot}P_i}]}{[\mathrm{M{\cdot}ATP}]} = 9,$$

$$K'_{(\mathrm{pH}=8)} = \frac{[\mathrm{ADP}][\mathrm{P_i}]}{[\mathrm{ATP}]} \simeq 10^7 \text{ M}.$$

For a discussion of these values see Trentham, Eccleston, and Bagshaw (1976, p.230) and references therein.

With reference to the lactate dehydrogenase-catalysed reaction, it will be seen that

$$K' = K_1 K^{\mathrm{B}} K_2.$$

K_1 and K_2 are themselves in this case the products of two constants each (the association of pyruvate and H^+ is taken as a single coupled equilibrium; see Chapter 5). The ratio $K_1/K_2 = 10^4$ is in agreement both with the steady state enzyme kinetic measurements and the difference between the 'on-enzyme' and the overall equilibrium constants.

An obvious but interesting point arises from the above reasoning. The direction of the lactate dehydrogenase-catalysed reaction demands that K_1 is written as an *association* constant and K_2 as a *dissociation* constant. If the overall Gibbs energy change of a reaction is small, as for instance in the hydrolysis of glucose 6-phosphate, the affinity (specificity) for substrate cannot be large without making the other equilibria of the sequence unfavourable. It seems possible that the linking of some processes to the hydrolysis of ATP or GTP may serve the purpose of increasing the overall Gibbs energy change and thus increasing the specificity of the reaction. This may be the case, for instance, in the steps between the formation of aminoacyl-tRNA and the elongation of polypeptide chains in ribosomes.

Returning to the reaction of ATP with myosin, it is clear that in this process – as well as in the case of transport ATPases – it is of particular interest to define the steps of energy transduction. While the data for the Gibbs energy changes obtained from transient kinetic and binding studies summarized by Gutfreund and Trentham (1975) will need revision, they do give a qualitative picture. Equilibrium measurements, together with the calorimetric investigations of Woledge and his colleagues (Woledge, 1977, Kodama and Woledge, 1979) allow one to draw up a table which provides a preliminary picture of Gibbs energy, enthalpy, and entropy changes during the steps of ATP hydrolysis by myosin. Much remains to be done before a satisfactory explanation of the transduction of chemical into mechanical energy can be claimed. During muscle contraction the steps listed in Table 4.5 are linked to a complex sequence of processes involving the interaction between myosin and actin fibres as well as with proteins which act as receptors for calcium stimulation and control (see for instance *Cold Spring Harbor Symposium on Quantitative Biology,* **37**, 1973; Adelstein and Eisenberg, 1980). It would be very interesting to discuss these problems in greater detail and to define some of the outstanding problems. However, this would require a description of the components of muscle fibres which is beyond the scope of this book.

One interesting consequence of the relation between substrate and product binding and the 'on-enzyme' equilibrium can be illustrated with myosin ATPase. Mannherz, Schenck and Goody (1974) demonstrated the formation of protein-bound ATP when the ATPase was mixed at high concentration with ADP, Mg^{2+}, and inorganic phosphate. From the equilibrium data one could conclude that the energy for ATP production could be provided by a process which causes the dissociation of this product from the protein. This is an energy-requiring process involving a conformation change in the protein. There have been some speculations that this could be the role of the proton gradient in mitochondrial ATP synthesis.

Table 4.5 Thermodynamic quantities relating to the hydrolysis of ATP by myosin

Reaction step	K'	ΔG/ kJ mol^{-1}	ΔH/ kJ mol^{-1}
Myosin			
⇅	4.5×10^3	−20.8	
Myosin–ATP			
⇅	1.8×10^9	−52.8	−90
*Myosin–ATP			
⇅	9	− 5.4	+83
**Myosin–ADP–P_i			
⇅	15	− 6.7	
*Myosin–ADP–P_i			−88
⇅	8×10^{-3}	+12.0	
*Myosin–ADP			
⇅	3.5×10^{-3}	+14.0	
Myosin–ADP			+72
⇅	2.7×10^{-4}	+20.4	
Myosin			

The asterisks against myosin indicate the degree of protein fluorescence enhancement. The equilibrium constants refer to pH 8 at 21 °C. The product of all the individual equilibrium constants $K'_{overall} = 8.27 \times 10^6$, which corresponds to $\Delta G' = -39$ kJ mol^{-1} for the hydrolysis of ATP. Considering the complexity of the system and the range of individual equilibrium constants reported, this is in reasonable agreement with $\Delta G^{\circ\prime} = -33$ kJ mol^{-1} reported for the hydrolysis of ATP by Rosing and Slater (1972). For detailed references to the determinations of the equilibrium constants see Trentham *et al.* (1976, p. 230) and for the calorimetric measurements of ΔH values see Woledge (1977).

4.10 Le Chatelier's principle and the effects of temperature and pressure on equilibria

If we alter a variable the system will adjust in the direction that minimizes the applied change in that direction. For instance, if the system is in equilibrium between two states A and B and the process A → B has a negative ΔV, an increase in pressure will shift the equilibrium towards B to decrease the volume. Similarly, if B → A has a positive ΔH, i.e. heat is adsorbed, a rise in temperature will shift the equilibrium in that direction. The principle of Le Chatelier (1888), expressed in its most general form, states that 'a system, when subjected to a perturbation, responds in a way that tends to eliminate its effect'. This argument can be applied to the response of an elastic fibre to the application of tension, to the addition of a reactant to a system in chemical equilibrium, and to many other phenomena involving the application of some physical or chemical change. In this section a discussion of the responses of systems at equilibrium to changes in temperature and pressure will be used to illustrate how information about the thermodynamic properties can be obtained from studies of responses to these particular variables. We are giving here further applications of the theory discussed in Chapter 3.

From equations (3.33) and (3.32) it follows that

$$\left(\frac{\partial G^\circ}{\partial p}\right)_T = V, \quad \left(\frac{\partial \Delta G^\circ}{\partial p}\right)_T = \Delta V \tag{4.21}$$

and

$$\left(\frac{\partial G^\circ}{\partial T}\right)_p = -S, \quad \left(\frac{\partial \Delta G^\circ}{\partial T}\right)_p = -\Delta S.$$

From the above and the expression $\Delta G = \Delta H - T\Delta S$ one obtains the Gibbs–Helmholtz relation between the enthalpy and the temperature dependence of the Gibbs energy:

$$\left(\frac{\partial \Delta G/T}{\partial T}\right) = -\frac{\Delta H}{T^2}.$$

For the derivation of the van't Hoff equation (4.22) one uses the relation $\Delta G^\circ/T = -R \ln K$:

$$-R\left(\frac{\partial(\ln K)}{\partial T}\right)_p = -\frac{\Delta G^\circ}{T^2} + \frac{1}{T}\left(\frac{\partial \Delta G^\circ}{\partial T}\right)_p$$

$$= -\frac{\Delta H^\circ}{T^2} + \frac{\Delta S^\circ}{T} + \frac{1}{T}(-\Delta S^\circ)$$

and thus

$$-R\left(\frac{\partial(\ln K)}{\partial T}\right)_p = -\frac{\Delta H^\circ}{T^2}. \tag{4.22}$$

Noting that

$$\left(\frac{\partial(\ln K)}{\partial T}\right)_p = \left(\frac{\partial(\ln K)}{\partial(1/T)}\right)_p \frac{d(1/T)}{dT} = -\frac{1}{T^2}\left(\frac{\partial(\ln K)}{\partial(1/T)}\right)_p, \tag{4.23}$$

it follows that

$$\left(\frac{\partial(\ln K)}{\partial(1/T)}\right)_p = -T^2\left(\frac{\partial(\ln K)}{\partial T}\right)_p = -\frac{\Delta H^\circ}{R}. \tag{4.24}$$

The linear relationships given by equations (4.24) mean that the slope of a plot of ln K versus T^{-1} at constant pressure (called a van't Hoff plot) will be equal to $-\Delta H°/R$. It must be noted that such a plot will be linear *if and only if* $\Delta H°$ is independent of T. Since $(\partial \Delta H/\partial T)_p = \Delta C_p$, this can be true only if $\Delta C_p = 0$.

Examples of such plots are discussed in Section 4.3 for the equilibria of reactions catalysed by alcohol dehydrogenase. In this case $\Delta H°$ was found to be essentially independent of temperature over the range investigated.

If $\Delta H°$ is temperature dependent then, of course, the slope of such a plot will vary continuously with temperature and thus indicates a finite value for ΔC_p. For a more detailed discussion and examples of changes in heat capacity see Section 6.2.

From equation (4.21) we can deduce that the equilibrium constant of a process will vary with pressure according to the relation

$$\left(\frac{\partial(\ln K)}{\partial p}\right)_T = -\left(\frac{\partial \Delta G°/RT}{\partial p}\right)_T = -\Delta V/RT. \tag{4.25}$$

A plot of ln K versus p will result in a straight line if and only if ΔV is constant over the range of pressures used in the experiment. This condition only appertains if the differential compressibility $\Delta \kappa = (\partial \Delta V/\partial p)_T$ is zero. Otherwise curvilinear plots will be obtained.

In the exploration of different contributions to the position of an equilibrium of a complex process it is very instructive to compare the magnitudes of the effects of temperature and pressure. Such information is, of course, also important for the design of experiments depending on the perturbation of equilibria by changes of intensive parameters. It must be pointed out that there are interesting cross-effects when temperature and pressure changes are studied. For example, the studies of Engelborghs, Heremans, De Mayer, and Hoebeke (1976) on the effects of these parameters on the assembly equilibrium of tubulin protomers show that the magnitude of the volume change is temperature dependent. In some systems a 20 °K change in temperature can even result in a change in sign of the volume change.

Table 4.6 gives the values for molar volume changes of a number of processes of biological interest. Considerable interest in the effects of pressure on quite complex biological systems has a long history (see, for instance, Cattell, 1936) and quite extensive documentation is found in a symposium (*Symposia of the Society for Experimental Biology*, **26**, 1972) on the effects of pressure on organisms.

For small changes in pressure, equation (4.25) approximates to

$$\Delta K/K = -\Delta p \, \Delta V/RT.$$

At 298 K, $RT = 24.4 \times 10^3$ cm^3 atm mol^{-1} and it follows that a reaction with a volume change $\Delta V = 25$ cm^3 mol^{-1} subjected to a pressure change of 100 atm will change its equilibrium constant by 10% ($\Delta K/K = 0.1$). For example, phosphate buffer at a pH near neutrality is perturbed by approximately 0.1 pH unit per 100 atm pressure change. From

Table 4.6 Volume changes associated with some representative processes (at 25 °C unless otherwise stated)

	$\Delta V/cm^3\ mol^{-1}$
Ionization of buffers	
Phosphate (p*K* 2.1)	−15.7
Phosphate (p*K* 7.0)	−24.0
Acetate	−10.7
Bicarbonate	−25.4
Carbonate	−25.6
Borate	−32.1
Tris	+ 1.0
$MgATP^{2-} \rightleftharpoons Mg^{2+} + ATP^{4-}$	−23
Hydrophobic effects	
CH_4 in C_4H_{14} : CH_4 in H_2O	−22.7
C_2H_4 in C_4H_{14} : C_2H_4 in H_2O	−18.2
Hydrogen bonding	
α-Helical poly-L-glutamate: random coil	+ 1.0
Poly(A + U) nucleotide helix: random coil	− 1.0
Protein self-assembly per monomer	
Myosin filament propagation	−280
Tubulin propagation at 35 °C	−26
Tubulin propagation at 15 °C	−50
Protein denaturation	
Chymotrypsin	−43
Myoglobin	−100

Table 4.6 it is apparent that with increasing pressure the pH will shift so as to increase the ratio of secondary to primary phosphate (since ΔV is negative when this shift takes place) and thus pH will increase.

For comparison of the magnitudes of temperature changes on the amplitudes of $\Delta K/K$ the simplified equation

$$\Delta K/K = \Delta H \Delta T/RT^2$$

can be used. The enthalpy of ionization of many groups with p*K* values in the neutral and alkaline region is around 45 kJ mol^{-1}. However, the ionizations of phosphate and of carboxylate groups are characterized by $\Delta H \simeq 0$. Using $\Delta H = 45$ kJ mol^{-1} as an example we obtain

$$\Delta K/K = 0.06\ K^{-1},$$

which corresponds to a 6% change in the equilibrium constant for 1 K change in temperature.

The combined use of temperature and pressure as variables in experiments designed to obtain precise values for ΔV can be illustrated by the following

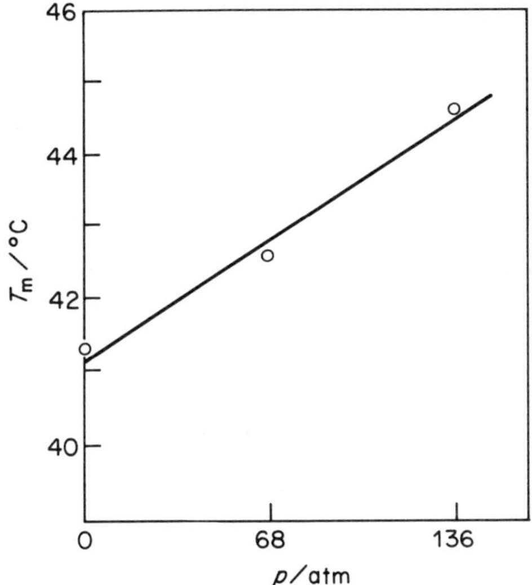

Figure 4.7 The pressure dependence of the melting temperature of dipalmitoylphosphatidylcholine (DPPC). (From data provided by Mountcastle, Biltonen and Halsey, 1978.)

example. Consider the reversible thermal transition of a biopolymer:

$$P_N \rightleftharpoons P_D.$$

The temperature at which $\ln K = \ln ([P_D]/[P_N]) = 0$ is referred to as the melting temperature (see detailed discussion in Chapter 6). At this temperature (T_m),

$$\Delta G^\circ = 0 \quad \text{and} \quad T_m = \Delta H/\Delta S$$

or

$$T_m = \frac{\Delta U}{\Delta S} + \frac{P\Delta V}{\Delta S}\,.$$

We can employ the Clausius–Clapeyron equation, in the form given in (3.27a), to calculate the variation of T_m with pressure. Rearranging that equation we obtain

$$\Delta V = \frac{\Delta H_m}{T_m}\,\frac{dT_m}{dp}\,.$$

We note that equation (3.27a) was applied to a system of a pure single component. Here we are dealing with a system of at least two components – the biopolymer and the surrounding solvent. However, the composition of both components is fixed, and does not change during the melting process, so that we can apply (3.27a) to this system.

Thus, if ΔH is known for the reaction, ΔV can be determined experimentally by evaluation of the dependence of T_m on pressure. An example of such evaluation is provided by the melting of a phospholipid membrane, dipalmitoylphosphatidylcholine (DPPC), as shown in Figure 4.7. In this system $T_m = 313$ K at 1 atm pressure, $\Delta H^\circ = 6$ kcal mol^{-1} of phospholipid. Since $\Delta T_m/\Delta P = 0.04$ K atm^{-1}, ΔV was calculated to be 26 cm^3 mol^{-1}, a value in good agreement with the directly determined ΔV for the transition.

Clearly a detailed knowledge of the temperature and pressure dependence is valuable, not only for the derivation of thermodynamic parameters, but also for the characterization of complex biochemical processes in terms of different ionic and other contributions.

CHAPTER 5

Ligand-binding equilibria

5.1 Introduction

The interaction of small molecules with macromolecules of biological systems and with specific receptor sites on surfaces of supramolecular organizations is one of the most extensively studied phenomena in biochemical research. The subject includes a vast range of important biochemical phenomena: the binding and release of protons by basic and acidic groups in proteins; the reversible binding of oxygen, carbon monoxide, and other compounds by myoglobins and haemoglobins; the non-covalent association of serum albumin with fatty acid anions and other compounds containing non-polar groups; the association of small cations, especially calcium and magnesium ions, with proteins and nucleic acids. These reversible interactions are closely related to the combination of enzymes with substrates and inhibitors; the kinetic phenomena of enzyme–substrate interaction commonly involve reversible as well as irreversible processes. Antigen–antibody reactions, with their high specificity, furnish other important examples; in this case both reactants are macromolecules, but the principles involved are not essentially different.

The fact that binding processes can frequently be coupled to associated structure changes in the protein or the ligand or both has interesting energetic and functional consequences. Ligand-induced conformation changes are often an essential part of the formation of the catalytically active complex of enzymes or of the transduction of information or energy. The energy changes involved in these important phenomena are usually due to non-covalent forces and are small compared with those accompanying formation or breakage of covalent bonds. However, binding equilibria can be measured sufficiently accurately to discriminate between processes which differ only in small energy increments. In fact the discrimination in terms of energy changes is much finer than any explanation in terms of structure and mechanism. Such studies are, therefore, an essential part of the physicochemical characterization of many biological phenomena ranging from enzyme catalysis and its control, hormone action, membrane transport, and nerve conduction to muscle contraction and other forms of motility.

Two points of nomenclature must be mentioned first. In the general (inorganic) literature of metal ion binding the molecule with the metal binding site (for instance ethylenediaminetetraacetate, EDTA) is called the ligand. In discussions of ions or other small molecules binding to macromolecules, the small molecules are called the ligands. We follow the latter practice here.

Ionic equilibria are always described by convention using *dissociation* constants. An example is the equilibrium

$$ROH \rightarrow RO^- + H^+,$$

for which

$$K = [RO^-][H^+]/[ROH].$$

In all other cases we employ *association* constants to describe binding equilibria. For instance, the association of myoglobin with oxygen is defined by

$$Mb + O_2 \rightleftharpoons MbO_2$$

and has the association constant

$$K = [MbO_2]/[Mb][O_2].$$

Clearly there is an inverse relation between dissociation and association constants. The sign of the standard Gibbs energy is reversed for association as compared to dissociation. Using subscripts D and A for dissociation and association we obtain

$$\Delta G^\circ_D = -RT \ln K_D = RT \ln (1/K_D) = RT \ln K_A = -\Delta G^\circ_A,$$

as would be expected for a change in direction of the reaction.

Several problems arise when steady state enzyme kinetics are used, as described by the classical Michaelis–Menten scheme, to relate the fraction of enzyme sites occupied by substrate. While it is true that in some systems the interconversion of enzyme–substrate complex to enzyme–product complex is much slower than any other reaction step, and that in the scheme

$$E + S \rightleftharpoons ES \rightarrow E + P$$

the Michaelis constant K_m can be regarded as the dissociation constant

$$K_m = [E][S]/[ES],$$

this assumption is often not justified (for detailed discussion see Gutfreund, 1975, pp. 134–135). When the Michaelis equilibrium assumption is made, the equation for the ratio of the rate v at a specified substrate concentration to the maximum rate V at saturating substrate concentration is defined by

$$\frac{v}{V} = \frac{[S]}{[S] + K_m}.$$

The data may fit an equation of the same form, even when K_m is not a true dissociation constant. The ratio v/V has to be used with caution as a measure of fractional saturation and this should only be done if there is no direct method available for the study of the true equilibrium for substrate binding.

Unfortunately the form of the Michaelis constant in the classical literature (as a dissociation constant) also resulted in the true substrate binding constants K_s being expressed as dissociation constants in the literature. The above problems arise in connection with studies of the cooperativity of substrate binding to enzymes with several binding sites, as we shall see later.

The analysis of ligand binding should first be attempted by seeing whether the data can fit equations for identical and independent (non-interacting) sites. Only if there is serious doubt about an adequate fit to such equations is one justified in attempting to fit the data to equations describing more complex mechanisms. Different experimental methods for the study of ligand binding provide different primary information. Equilibrium dialysis, for instance, allows determination of free ligand in the dialysate and total ligand (free and bound) in the protein solution. It is generally justifiable to assume that the concentration of free ligand is the same in both solutions. Titration of ion binding sites in the presence of pH electrodes or other ion-specific probes provides a plot of total ion added against free ion concentration. Spectrophotometric titration of proteins with suitable ligands provides plots of fractional saturation against total ligand added. In each of these types of experiment one can, of course, calculate for each total concentration of ligand the corresponding concentrations of free and bound ligand. Different equations are, however, commonly used in the study of ligand binding in different systems.

Another practical consideration is whether accurate determination of the number of binding sites or of the association constants is the primary objective. This can be illustrated with the simulation of the titration of a protein with a ligand, which results in some physical signal (spectroscopic, electrode measurement, etc.) due to complex formation. If the experiment is carried out in the differential mode, ligand being added simultaneously to the protein solution and to a buffer solution, the difference between the signal from the two will result in curves of the type shown in Figure 5.1. At high protein concentrations (concentration of binding sites in excess of reciprocal association constant) accurate values for the total number of sites are obtained. More accurate values for association constants are, however, obtained at lower protein concentrations, using methods for the evaluation described below. The choice of protein concentration must, of course, be compatible with the sensitivity of the measurement.

5.2 Linear methods for the evaluation of stoicheiometry and association constants

Consider a solution containing a ligand, and a molecule that contains n sites capable of binding the ligand. The ligand-binding molecule is usually, but not

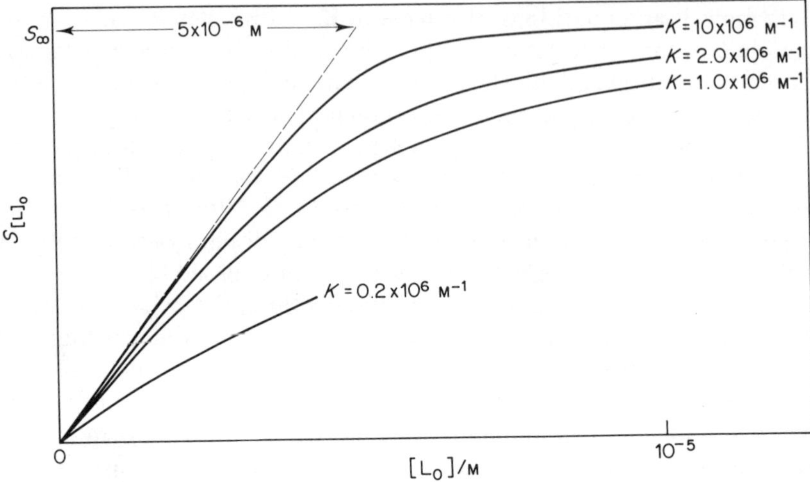

Figure 5.1 Simulations of differential titrations of protein solutions containing $[B]_o = 5 \times 10^{-6}$ M binding sites for ligand L. The total ligand concentration, free and bound, is $[L]_o$ and the differential signal between protein solution and reference solution containing the same amount of ligand is $S_{[L]_o}$ at each ligand concentration. The proteins in the different solutions have sites with different association constants as indicated on the traces

necessarily, a macromolecule. As the concentration of ligand in solution is increased, the average number of ligand molecules bound per macromolecule increases from zero, in the absence of ligand, to n, at complete saturation of binding sites. At any given concentration of free ligand, the binding that we measure is this average value, which must lie between zero and n. We denote this number by the symbol $\bar{v}$, the bar superscript being used to denote an average over all the ligand-binding molecules in the system. Alternatively we may prefer to consider the fraction of all binding sites in the system that is occupied by ligand. This fraction is denoted as y, and its value must lie between 0 and 1:

$$y = \bar{v}/n.$$

The fraction of unoccupied sites, i.e. the degree of dissociation, is obviously equal to $1 - y$, and will be denoted by θ.

In some cases we know the value of n. For the binding of oxygen by haemoglobin, for instance, $n = 4$. In other cases, such as the binding of a fatty acid anion by serum albumin, the value of n is less certain; there are some binding sites of high affinity, and others that are much weaker. We can, however, determine $\bar{v}$ at a series of values of free ligand concentration, and by appropriate methods of analysing the data we can arrive at estimates of the number of binding sites of the various classes.

In a classical paper on 'attractions of proteins for small molecules and ions', Scatchard (1949) proposed a still widely used method for treating data for protein–ligand interactions. He defined $\bar{v}$, the average number of ligands (L) associated with the protein (P) in the reaction

$$P + L \rightleftharpoons PL + L \rightleftharpoons PL_2 \ \ldots\ PL_{n-1} + L \rightleftharpoons PL_n$$

by the equation

$$\bar{\nu} = \sum_{\nu=0}^{n} [PL_\nu]\, \nu / \sum_{\nu=0}^{n} [PL_\nu], \tag{5.1}$$

where n is the total number of binding sites and ν the number of ligand molecules (from 0 to n) bound to each protein species. The denominator is equivalent to total protein concentration $[P]_0$ in all forms and the numerator gives the concentration of occupied sites:

$$\bar{\nu} = \frac{\text{concentration of bound ligand}}{\text{total concentration of protein}}$$

The law of mass action for binding ligand to a protein with a single binding site can be expressed by

$$K = [PL]/([P]_0 - [PL])[L], \tag{5.2}$$

where $[P]_0$ is the total protein concentration and [PL] and [L] are the concentrations of bound and free ligand respectively. Since $[PL]/[P]_0 = y$, we can write

$$y + yK[L] = K[L]. \tag{5.3}$$

If this is extended to the case of multiple but identical binding sites, we can use $\bar{\nu} = ny$ as defined by equation (5.1) and equation (5.3) becomes

$$\bar{\nu} + \bar{\nu}K[L] = nK[L]. \tag{5.4}$$

Equation (5.4) can be solved for [L] to give the original Scatchard (1949) equation

$$\bar{\nu}/[L] = nK - \bar{\nu}K, \tag{5.5}$$

which was developed as a test for the identity and independence of sites, as well as for estimating the number of binding sites per protein molecule. A plot of $\bar{\nu}/[L]$ as ordinate against $\bar{\nu}$ as abscissa gives an ordinate intercept Kn and abscissa intercept n. The linearity of such a plot tests the identity and independence of sites.

If the multiple binding sites on a macromolecule are independent, the ligand binding sites can be reproduced by a model system of monovalent molecules with the same set of K values. If there are j classes of such independent sites, each class characterized by a single K value, the binding curve is described by

$$\bar{\nu} = \sum_{i=1}^{j} \frac{n_i K_i [L]}{1 + K_i [L]}, \tag{5.6}$$

where n_i and K_i are the numbers and association constants of each species of binding site.

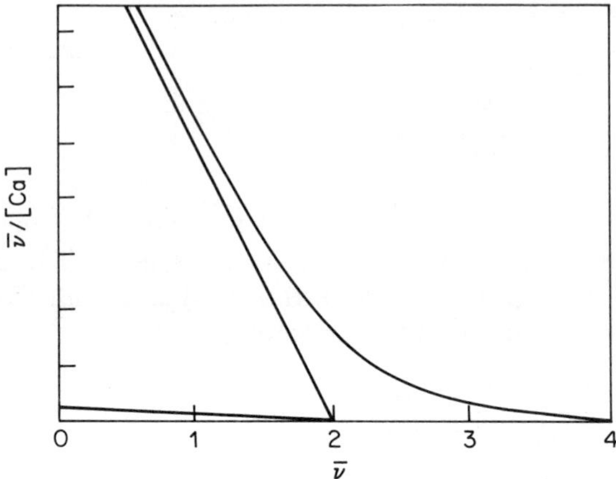

Figure 5.2 A Scatchard plot for binding of calcium to troponin C simulated from data of McCubbin and Kay (1973). The straight lines correspond to the slopes due to the two sets of sites referred to in the text. As pointed out in the text, numerical analysis, rather than fitting tangents to asymptotes, is needed for the evaluation of multiple association constants

Figure 5.2 illustrates a simulated binding experiment for the troponin–calcium system in terms of a Scatchard plot. McCubbin and Kay (1973) found that troponin C from skeletal muscle has two high affinity ($K = 2 \times 10^7$ M^{-1}) and two lower affinity ($K = 5 \times 10^5$ M^{-1}) binding sites for calcium, the latter being involved in the regulation of muscle contraction. It must be emphasized that association constants cannot be obtained directly from the asymptotes of the Scatchard plot obtained for a system with binding sites having two distinct affinities. This should be clear from inspection of Figures 5.2–5.4, and the reader is referred to the discussion by Norby, Ottolenghi, and Jensen (1980). The non-linear least squares method described in the appendix can be used to derive the association constants by fitting the data to equation (5.6).

Another example of clearly distinct binding sites is illustrated in Figure 5.3, where the analysis of data from binding studies of cytosine triphosphate to aspartyl transcarbamylase are described.

In systems with positive (cooperative) interactions (see Section 5.9) one cannot explain the data by the above model. In an anticooperative system (negative cooperativity) one generally gets the same type of Scatchard plot as for non-identical sites, and chemical evidence is needed to distinguish negative cooperativity from non-identical sites (see p. 184 for discussion of an anticooperative system).

A Scatchard plot for a group of equivalent and independent sites is a straight line, with a negative slope. If there is another group of sites, also equivalent and independent among themselves but with a different (smaller) binding constant, the line will become curved in the lower region as the $\bar{\nu}$ axis is approached, extending out to higher values as $\bar{\nu}$/[L] diminishes. This behaviour is

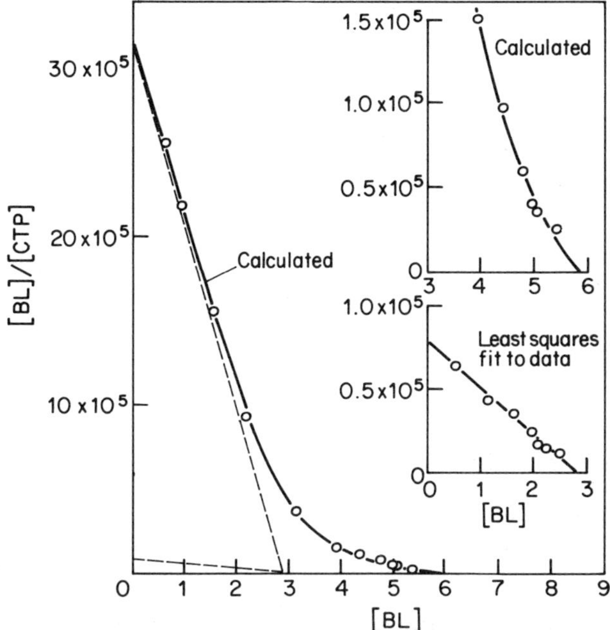

Figure 5.3 Binding studies of cytosine triphosphate (CTP) to aspartyl transcarbamylase reported by Winlund and Chamberlin (1970). The experiments were carried out by equilibrium dialysis at 4 °C in 0.01 M potassium phosphate pH 7.0, 2 mM 2-mercaptoethanol, and 0.2 mM sodium EDTA. Enzyme was present at 20 mg ml^{-1}

illustrated in Figure 5.4 from the work of Scatchard, Scheinberg, and Armstrong (1950) which shows the binding of thiocyanate ion to human serum albumin. The data at first give a rough fit to a straight line, corresponding to an intercept of about 10 750 on the ordinate axis. However, as thiocyanate concentration increases, the curve deviates from this line, indicating the presence of a second set of binding groups, weaker than the first set. The binding is affected by electrostatic effects, since each ion bound makes it more difficult to bind another ion of the same sign. These effects are taken into account in the calculation of the plotted data, as explained by the authors (see also Edsall and Wyman, 1958, pp. 645–651). The data can be fitted, within the limits of experimental uncertainty, by assuming two classes of groups: class 1, with 10 groups and $K_1 = 1000\ \text{M}^{-1}$; class 2, with 30 groups and $K_2 = 25\ \text{M}^{-1}$.

Later Scatchard, Coleman, and Shen (1957) and Scatchard, Wu, and Shen (1959) studied serum albumin that had been treated to remove the strongly bound fatty acid anions that are present in most preparations. This purified albumin had one binding site of very high affinity ($K_1 = 4.7 \times 10^4\ \text{M}^{-1}$) not observed in the earlier work, eight groups with $K_2 = 2000\ \text{M}^{-1}$, and 18 groups with $K_3 = 60\ \text{M}^{-1}$. The binding of chloride ions could be

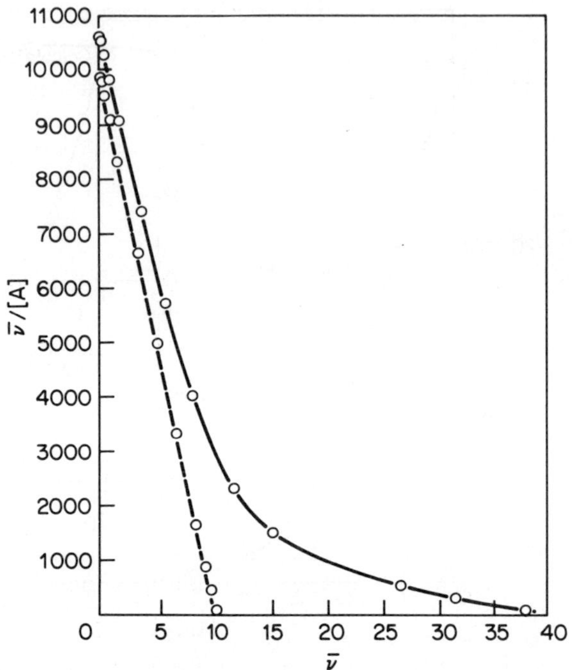

Figure 5.4 Scatchard plot for binding of thiocyanate ions to human serum albumin. (From Scatchard *et al.*, 1950.)

described by the same sets of groups, with the same ratio of K_1, K_2, and K_3 values, but with the absolute values of all the K_i diminished by a factor of 20.

Another approach to the problem of evaluating the number and properties of binding sites can be used when curves of the type shown in Figure 5.1 are obtained from spectrophotometric titration and the protein concentration is known. The method determines the total number of binding sites and so permits evaluation of the concentration of sites per protein molecule. The degree of liganding (or saturation of sites) y has values from 0 to 1. This can be evaluated at each total ligand concentration from the ratio of the signal $S_{[L]}$ at a given ligand concentration to the signal S_{max} at ligand saturation (where the signal gives the difference as indicated on p. 159):

$$y = S_{[L]}/S_{max} = [BL]/[B]_o$$

where [BL] is the concentration of liganded sites and $[B]_o$ the total number of binding sites. The law of mass action gives

$$K = [BL]/([B]_o - [BL])([L]_o - [BL]),$$

where $[L]_0$ is the total ligand added. But we know that

$$[BL] = y[B]_o$$

and therefore, by substitution, we obtain

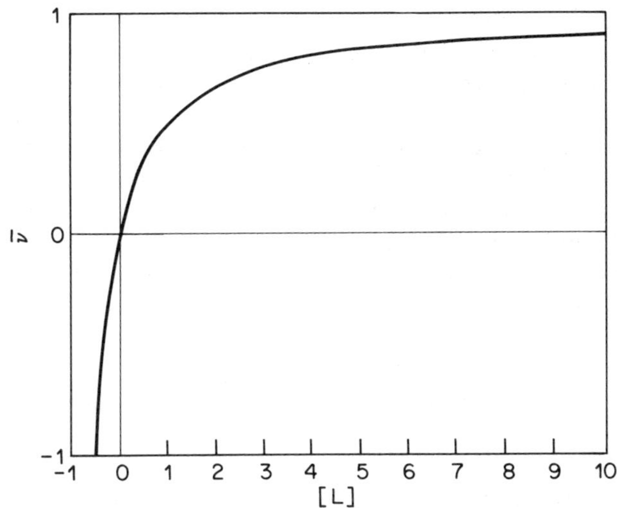

Figure 5.5 The relation between the binding curve and a rectangular hyperbola. The equation for a rectangular hyperbola is $xy = a$. One can transform the binding equation $\bar{\nu} = [L]/\{[L] + (1/K)\}$ into $\{[L] + (1/K)\}(\bar{\nu} - 1) = -1/K$. Then $[L] + (1/K) = x$, $\bar{\nu} - 1 = y$ and $-1/K = a$ in the above equation for a hyperbola. The association constant $K = 1$ and the number of binding sites (i.e. the maximum value for $\bar{\nu}$) is 1. The asymptotes of the hyperbola are $[L] + (1/K) = 0$, which gives $[L] = -1/K$, and $\bar{\nu} - 1 = 0$, which gives $\bar{\nu} = 1$. For more detail see Cornish-Bowden (1976, p. 32)

$$1/(1 - y) = K\,[L]_o/y - K[B]_o\,; \tag{5.7}$$

a plot of $1/(1 - y)$ as ordinate and $[L]_o/y$ as abscissa gives K as the slope and the intercept on the abscissa is $[B]_o$, the number of binding sites.

In the modern literature on saturation curves (plots of y against $[L]_o$) one often reads the comment that 'the data fit a rectangular hyperbola' as evidence for equivalence and independence of sites. The transformation of the equation $y = [L]/(K + [L])$ into the form $xy = a$ is shown in Figure 5.5. If the data fit a rectangular hyperbola, linear plots are obtained by the method of Scatchard and from equation (5.7). Section 5.8 is concerned with non-linear plots resulting from heterogeneity in binding sites and interactions between them.

5.3 Hydrogen ion dissociation

Acid–base equilibria are usually described by dissociation constants: for neutral acids

$$AH \rightleftharpoons A^- + H^+ \quad \text{and thus} \quad K = [A^-][H^+]/[AH]$$

and for cationic acids

$$AH^+ \rightleftharpoons A + H^+ \quad \text{and thus} \quad K = [A][H^+]/[AH^+].$$

The study of such equilibria is of great importance in the interpretation of energy changes and reactivities of many compounds involved in biological processes. Examples of the linkage between ionic dissociation and biochemical reactions are given in Section 4.4. In fact many fundamental contributions to our understanding of the behaviour of ionizing groups have come from studies which were primarily stimulated by problems arising in connection with physiological and biochemical investigations. Three texts (Cohn and Edsall, 1943; Edsall and Wyman, 1958; Tanford, 1961) provide advanced treatments of this subject matter. Hydrogen ion dissociation equilibria are usually characterized by the pK of an ionizing group. The relation

$$\mathrm{p}K = -\log K$$

is used in the derivation of the Henderson–Hasselbalch equation:

$$\mathrm{pH} = \mathrm{p}K + \log ([\mathrm{A}^-]/[\mathrm{AH}]).$$

The degree of dissociation θ ($0 \rightarrow 1$) is defined by

$$\theta = \frac{[\mathrm{A}^-]}{[\mathrm{A}^-] + [\mathrm{AH}]}$$

for a neutral acid and by

$$\theta = \frac{[\mathrm{A}]}{[\mathrm{A}] + [\mathrm{AH}^+]}$$

for a cationic acid, and for the Henderson–Hasselbalch equation we can write

$$\mathrm{pH} = \mathrm{p}K + \log \{\theta/(1 - \theta)\} \tag{5.7a}$$

From a titration curve as shown in Figure 5.6 for the protonation of tris(hydroxymethyl)aminomethane one can calculate points for a plot of log $\{\theta/(1 - \theta)\}$ against pH, which will intercept the abscissa (pH axis) at pH = pK, that is when $[\mathrm{A}] = [\mathrm{AH}^+]$ at half dissociation. Some of the other plots described above for the evaluation of ligand association constants can also be transformed to analyse ionic dissociation.

The pK can be directly related to the standard Gibbs energy of dissociation by

$$\mathrm{p}K = -\log K = \Delta G°/2.303\ RT.$$

At 25 °C, pK = $\Delta G°$/5.71 kJ mol^{-1}. This calculation gives some idea of the sensitivity of equilibrium measurements to small energy changes. A change of one unit in pK, one order of magnitude in equilibrium constant, corresponds to only 5.71 kJ mol^{-1} change in Gibbs energy. Small changes in distance and alteration of the local dielectric constant can cause large changes in electrostatic energy. One of the interesting features of large polyelectrolyte molecules, such as proteins or nucleic acids, is their ability to

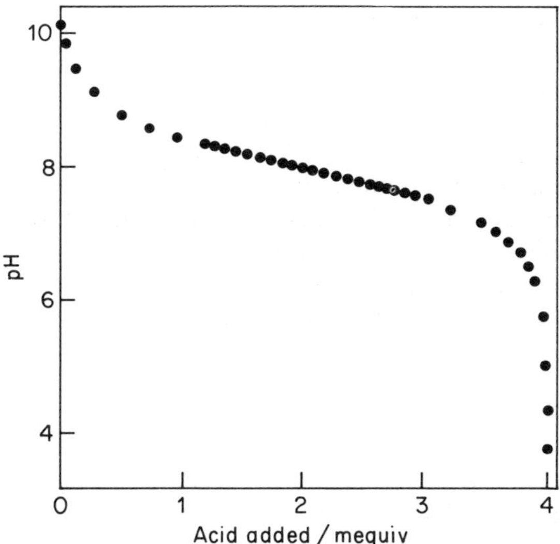

Figure 5.6 Titration curve for tris(hydroxymethyl)aminomethane (Tris). 40 ml of a solution containing 4 mequiv Tris was titrated with 1.0 N HCl at 25 °C. The solution was 0.6 M with respect to KCl. (From Bernhard, 1956.)

modulate local conditions which affect the properties of the many different ionizing groups which they contain.

Some of these points can be illustrated with data for smaller compounds with ionizing groups. For instance the effect of neighbouring charged groups on the pK values of carboxylic acids can be seen from the following comparison. The pK values refer to the dissociation of the proton on the right-hand side:

acetic acid	CH_3COOH	pK = 4.75,
aminoacetic acid (glycine)	$^+H_3NCH_2COOH$	pK_2 = 2.35,
malonic acid	$HOOCCH_2COOH$	pK_1 = 2.82,
malonic acid anion	$^-OOCCH_2COOH$	pK_2 = 5.69.

The repulsive force between the $—NH_3^+$ ion and the proton associated with the carboxylic group of glycine decreases the absolute value for the Gibbs energy of proton binding to -13.5 kJ mol^{-1} from -27.3 kJ mol^{-1} for acetic acid. In the first ionization of malonic acid (pK_1 = 2.82) the neighbouring unionized carboxyl group strengthens the acidity of the group that ionizes first. Conversely, the negative charge that is present after ionization of the first group increases the binding energy and decreases the acidity of the proton on the other carboxylic group.

In the case of neutral acids, where charge separation occurs during dissociation, local dielectric constants have a considerable effect, as can be predicted from the equation for the force between two charges e_1 and e_2 separated by the distance r (Coulomb's law),

$$F = ke_1e_2/r^2D,$$

and the work done in separating the charges from distance r_1 to r_2 is

$$W = \int_{r_1}^{r_2} F \, dr = \frac{ke_1e_2}{D}\left(\frac{1}{r_1} - \frac{1}{r_2}\right), \tag{5.8}$$

where D is the dielectric constant. In SI units the constant $k = 9 \times 10^9$ N m^2 C^{-2} and the magnitude of the charge on an electron or proton is 1.6×10^{-19} C (see Chapter 2, p. 14). When the charges are of opposite sign the force and work are negative and correspond to an attraction. The work required to separate two ions of opposite charge from a centre to centre distance of 0.3 nm to infinity in a vacuum ($D = 1$) would be

$$[9 \times 10^9 \times (1.6 \times 10^{-19})^2 \times 6.03 \times 10^{23}]/3 \times 10^{-10} = 463 \text{ kJ mol}^{-1},$$

which is a very large binding energy. However, in an aqueous solution, with the dielectric constant of water about 80, this binding energy would be reduced to about 5.8 kJ mol^{-1}, which is only about four times the kinetic energy of the molecules.

The two effects of neighbouring charges and of dielectric constants have to be taken into account when dissociation constants of ionizing groups and the energy of binding of ions to charged groups on surfaces of macromolecules are considered. Clearly, the different non-polar and charged groups, which make up the structures of protein molecules, can offer a wide range of environments. This environmental factor can be easily modulated by conformation changes, and often ligand binding will cause conformation changes by alteration of the local forces due to the resulting changes in environment.

Figure 5.7 shows the effects of changes in dielectric constant, produced by the addition of non-polar solvents to water, on the pK of acetate. Similar effects of increasing ionic strength on the pK values of neutral acids are to be expected. For instance the second pK of phosphate is reduced from 7.07 at ionic strength $I = 0.01$ to 6.67 at $I = 0.40$. It is of interest to record that, in some proteins, very large perturbations of pK values of particular groups have been found. For instance Schmidt and Westheimer (1971) have reported that an ε-lysine group at the active site of acetoacetate decarboxylase has a pK of 6.0, which is 4.7 units lower than that for free lysine (see also discussion in Section 5.8).

The heat of ionization can be determined either by titration in a calorimeter (Chapter 6) or from the temperature dependence of the pK. The van't Hoff equation (see p.152) can be written as follows in terms of pK values at two different temperatures:

$$\Delta H = (\text{p}K_{T_1} - \text{p}K_{T_2}) \frac{T_1T_2}{T_2 - T_1} \times 19.14, \tag{5.9}$$

which gives the heat of ionization in joules per mole. Typically the heats of ionization of carboxylic groups are very small (less than 5 kJ mol^{-1}) while

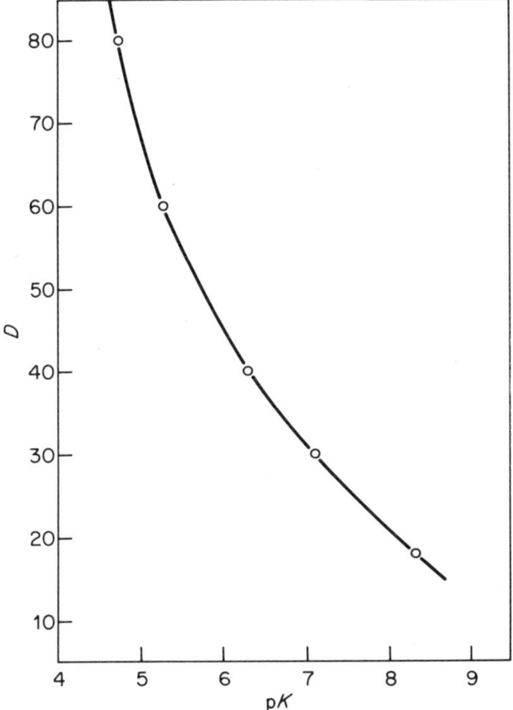

Figure 5.7 The effect of dielectric constant on the pK of acetate

those associated with —OH, —NH_3^+, and imidazole groups are between 30 and 50 kJ mol^{-1}. Like the Gibbs energy, the heat of dissociation can vary considerably with the environment, but the variation is not quite as large for the latter as has been found for the former. Table 5.1 gives some examples of heats of ionization of groups of biochemical interest.

5.4 Multiple binding sites: the statistical factor

Suppose HAH to have two ionizing groups which are in every way identical and which are separated by a sufficient distance so that there is no charge interaction. Titration of such a dibasic acid with a strong base would result in a curve which represents a single dissociation constant, but two equivalents of base will be needed to reach $\theta = 1$. In this case, in fact, the form of the titration curve would be the same, whether one equivalent of HAH or two equivalents of AH were titrated. However, it is of interest to examine the relation between the two notional dissociation constants:

$$K_1 = \frac{[\mathrm{HA}^-][\mathrm{H}^+]}{[\mathrm{HAH}]} \quad \text{and} \quad K_2 = \frac{[^-\mathrm{A}^-][\mathrm{H}^+]}{[\mathrm{HA}^-]} .$$

Table 5.1 Thermodynamic functions of some ionizing groups. All data are at 25 °C

Substance	pK	$\Delta G°$/kJ mol^{-1}	$\Delta H°$/kJ mol^{-1}	Reference
Acetic acid	4.756	27.14	−0.385	
Lactic acid	3.860	22.04	−0.414	
Succinic acid	4.207	24.02	3.188	
	5.636	32.19	−0.452	
Phosphoric acid	2.148	12.26	−7.648	
	7.198	41.10	4.109	
Ammonium ion	9.245	52.78	52.216	
Tris(hydroxymethyl)aminomethane	8.075	46.10	45.606	
Glycine	2.350	13.41	4.837	
	9.780	55.81	55.815	
Carbonic acid	6.352	36.26	9.372	
	10.329	58.96	15.075	
Amino acid side chains and related groups				
Lysine pK_3 (ε-ammonium)	10.79			
Lysyllysine (ε-ammonium)	10.05			
	11.01			
Arginine pK_3 (guanidinium)	12.5			

Tyrosine pK_3 (hydroxyl)	10.0				Martin, Edsall Wetlaufer, and Hollingworth (1958)
Glycylglycine	3.148 8.252	17.96 47.09		3.62 44.3	Smith and Smith (1942)
Ala-Ala-Ala-Ala	3.42 7.94				Ellenbogen (1952)
Carbobenzoxy-Pro-His-Gly (imidazole)	6.42	36.64		33	Koltun, Clark, Dexter, Katsoyannis and Gurd (1959)
N-Acetyl-L-isoasparagine	4.08				Nozaki and Tanford (1967)
N-Acetyl-L-isoglutamine	4.50				Nozaki and Tanford (1967)
Methylthioglycolate	7.8† 9.5†				Edsall and Wyman (1958, pp. 496–508)
Cysteine	8.3† 10.8†				Benesch and Benesch (1955)

For a more detailed discussion and further references see Edsall and Wyman (1958, pp. 496–508).
†These are hybrid constants each involving both the —NH_3^+ and the —SH groups.

Since, by definition, ^{-}AH and HA^{-} are indistinguishable, the above should be written in the form

$$K_1 = \frac{([HA^-] + [^-AH])[H^+]}{[HAH]} \quad \text{and} \quad K_2 = \frac{[^-A^-][H^+]}{[HA^-] + [^-AH]} .$$

According to the scheme

$$\begin{array}{ccccc} & & HAH & & \\ & k_a \nearrow & & \searrow k_c & \\ HA^- & & & & ^-AH \\ & k_b \searrow & & \nearrow k_d & \\ & & ^-A^- & & \end{array}$$

with four identical individual dissociation constants $k_a = k_b = k_c = k_d = k_o$, we have

$$k_a = \frac{[HA^-][H^+]}{[HAH]} \quad \text{and} \quad k_c = \frac{[^-AH][H^+]}{[HAH]} .$$

Thus $K_1 = k_a + k_c = 2k_o$ and similarly $1/K_2 = 1/k_b + 1/k_d$ and $K_2 = \frac{1}{2} k_o$ and $K_1 = 4K_2$. Although, in a system as defined here, the intrinsic constants cannot be directly determined experimentally, their physical significance is of interest. As will be seen, there are systems with identical ligand binding sites where the distribution of individual species P, PL_1, PL_2, etc., can be evaluated (see Section 5.5). From the kinetic scheme defined by $k_o = k_1/k_{-1}$ one can derive that at equilibrium

$$\frac{d([HA^-] + [^-AH])}{dt} = 2k_1[HAH] - k_{-1}([HA^-] + [^-AH])[H^+] = 0,$$

$$\frac{d[^-A^-]}{dt} = k_1([HA^-] + [^-AH]) - 2k_{-1}[^-A^-][H^+] = 0,$$

$$K_1 = 2k_1/k_{-1},$$

$$K_2 = k_1/2k_{-1}.$$

There are two ways for protons to dissociate from HAH and only one way for association to ^{-}AH or HA^{-}. In the next step there is only one way for protons to dissociate from ^{-}AH or HA^{-}, but two ways for protons to associate to $^{-}A^{-}$. It must be emphasized again that chemically ^{-}AH and HA^{-} are indistinguishable, although they can be formed in two ways from HAH. Continuing the same argument for a four-step dissociation,

$$AH_4 \rightleftharpoons {}^{-}AH_3 \rightleftharpoons {}^{2-}AH_2 \rightleftharpoons {}^{3-}AH_1 \rightleftharpoons {}^{4-}A,$$

thus

$$K_1 : K_2 : K_3 : K_4 = \frac{4}{1}\ \frac{3}{2}\ \frac{2}{3}\ \frac{1}{4}, \qquad \text{i.e.} \quad K_1/K_4 = 16,$$

and the ith dissociation constant K_i in a sequence of n steps is derived as follows:

$$K_1 = \frac{n}{1}\ k_o, \quad K_2 = \frac{n-1}{2}\ k_o, \quad \dots, \quad K_n = \frac{1}{n}\ k_o.$$

Clearly the overall equilibrium constant is given by

$$K = \prod_{i=1}^{n} K_i$$

and

$$K_i = k_o(n - i + 1)/i. \tag{5.10}$$

K_i is the macroscopic dissociation constant, while k_o is the microscopic (intrinsic) dissociation constant which gives the exact measure of the affinity for the individual binding process.

The importance of the dissociation constants for the individual steps (K_1 to K_n) becomes apparent when one wishes to calculate the concentrations of the different liganded species during the course of titration. This problem is of interest both in systems of identical and independent sites and for the more complex models discussed in Sections 5.7 and 5.8.

5.5 Intermediates during titration of multiple sites

For a molecule with four identical and independent sites the fractions which are unliganded, mono-, di-, tri-, and tetraliganded are illustrated in Figure 5.13 (see p. 194). In a system of this kind,

$$P \overset{K_1}{\rightleftharpoons} PL_1 \overset{K_2}{\rightleftharpoons} PL_2 \overset{K_3}{\rightleftharpoons} PL_3 \overset{K_4}{\rightleftharpoons} PL_4.$$

The total concentration of P is given by

$$[P] + [PL_1] + [PL_2] + [PL_3] + [PL_4] = 1 \qquad \text{(normalized)}.$$

Thus the total site concentration in this normalized form of the equation, where we take the value $y = 1$, defined by Scatchard as the average number of sites occupied per macromolecule, is 4:

$$y = \frac{\text{sites occupied}}{\text{total sites}}$$

$$= \frac{\tfrac{1}{4}[PL_1] + \tfrac{2}{4}[PL_2] + \tfrac{3}{4}[PL_3] + [PL_4]}{[P]_o}$$

and

$$\bar{v} = \frac{\text{moles of bound L}}{\text{total moles of protein}} = \frac{[PL_1] + 2[PL_2] + 3[PL_3] + 4[PL_4]}{[P]_o}.$$

To evaluate the concentrations of the individual species at any specified concentration of free ligand concentration the four macroscopic association constants $K_1 = 4k$, $K_2 = 3k/2$, $K_3 = 2k/3$, and $K_4 = k/4$ are used:

$$\left.\begin{aligned} [PL_1] &= [P]K_1[L] &&= 4[P]k[L], \\ [PL_2] &= [PL_1]K_2[L] &&= \tfrac{3}{2}[PL_1]k[L] &&= 6[P]k^2[L]^2, \\ [PL_3] &= [PL_2]K_3[L] &&= \tfrac{2}{3}[PL_2]k[L] &&= 4[P]k^3[L]^3, \\ [PL_4] &= [PL_3]K_4[L] &&= \tfrac{1}{4}[PL_3]k[L] &&= [P]k^4[L]^4. \end{aligned}\right\} \tag{5.11}$$

It follows that the total concentration $[P]_o$ is given by

$$[P]_o = [P]\{1 + 4k[L] + 6k^2[L]^2 + 4k^3[L]^3 + k^4[L]^4\} = [P](1 + k[L])^4$$

and

$$[P] = [P]_o/(1 + k[L])^4.$$

Given [L], the concentrations of all species can be calculated. The expression for the degree of liganding is obtained for substitution into equation (5.11):

$$y = \frac{[PL_1] + 2[PL_2] + 3[PL_3] + 4[PL_4]}{4[P](1 + k[L])^4}$$

$$= \frac{4[P]k[L] + 12[P]k^2[L]^2 + 12[P]k^3[L]^3 + 4[P]k^4[L]^4}{4[P](1 + k[L])^4}$$

$$= \frac{k[L](1 + k[L])^3}{(1 + k[L])^4} = \frac{k[L]}{1 + k[L]}.$$

Numerical examples of calculation of the concentrations of the different liganded species should illustrate the points made above (see also comparison of intermediate concentrations in systems with independent and with cooperative sites on p. 194). If k is set as 10^6 M^{-1} and free ligand concentration $[L] = 10^{-6}$ M then $y = 0.5$, and if $[P]_o = 10^{-6}$ M we find that $[P] = 6.25 \times 10^{-8}$ M, $[PL_1] = 2.5 \times 10^{-7}$ M, $[PL_2] = 3.75 \times 10^{-7}$ M, $[PL_3] = 2.5 \times$

10^{-7} M, $[PL_4] = 6.25 \times 10^{-8}$ M, and total ligand concentration $[L]_o = 3.0 \times 10^{-6}$ M (see also Table 5.3 and Figure 5.13). Optical titration curves give, at each point, moles of bound ligand from total ligand bound at saturation ($n[P]_o$ = sites) multiplied by S/S_∞ and

$$[\text{total ligand added}] - n[P]_o\ S/S_\infty = [\text{free ligand}].$$

An example of a method for the investigation of the statistical distribution of liganded species is the study of NADH binding to heart lactate dehydrogenase. Holbrook (1972) carried out a detailed quantitative analysis of the non-linearity of protein fluroescence quenching by ligands through energy transfer (Förster, 1959). In the case of the interaction of NADH with the four binding sites of the tetrameric lactate dehydrogenase spectral and fluorescence changes of NADH permit the conclusion that nucleotide binding can be described by a single association constant. Yet, it has been found that the quenching of protein fluorescence is not a linear function of degree of liganding by NADH. Holbrook (1972) postulated that this non-linearity was due to the fact that the first molecule of NADH, binding to one subunit, not only quenched the tryptophan fluorescence of that unit but also, to a lesser and distance-dependent extent, quenched the fluorescence of the tryptophan residues of the other units of the tetramer. This phenomenon would result in a smaller decrement of fluorescence as liganding of the four sites progresses. Holbrook was able to show that, for a number of oligomeric dehydrogenases, the data could be fitted to the equation

$$F = [1 - y(1 - x)]^n,$$

where F is the fluorescence of the protein with n ligands bound, relative to the fluorescence of the unliganded protein. The factor x is specific for the protein and ligand and y is the degree of liganding. For pig heart lactate dehydrogenase and NADH, $x = 0.59$. This results in values for F ranging from 0.59 for liganding of the first site to $F = 0.12$ for fully liganded enzyme. Figure 5.8 shows a comparison between the protein fluorescence and the degree of liganding as a function of NADH concentration. This has been simulated for the expected concentrations of the unliganded and the four liganded species of pig heart lactate dehydrogenase. Holbrook (1972) has shown that these curves are in agreement with NADH binding to four identical and independent sites.

5.6 Two-step binding processes

Strictly thermodynamic or kinetic information can give no positive evidence for a particular chemical mechanism. It does, however, provide the framework into which any proposed mechanism must fit. It also serves as an important guide to further experiments and the conditions suitable for the application of direct chemical and spectroscopic probes. A number of different proposals

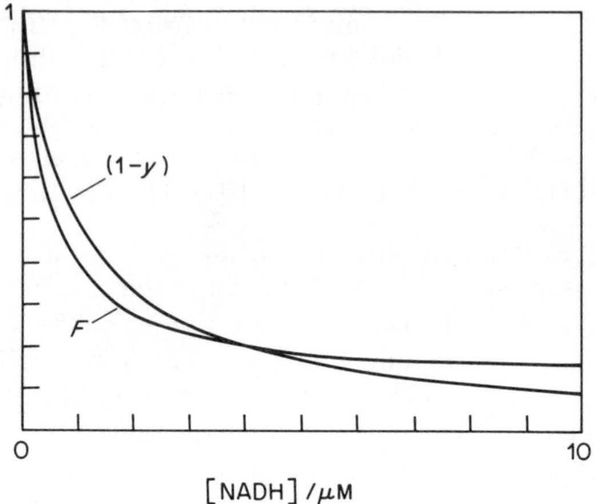

Figure 5.8 Simulation of degree of liganding, y, and protein fluorescence, F, for the titration of heart muscle lactate dehydrogenase with NADH. The association constant is 10^6 M^{-1}

have been made for the mechanism of ligand binding to macromolecules, especially proteins, in two distinct steps, when a bimolecular adsorption is followed by a monomolecular isomerization of the complex. In enzymes this isomerization of the substrate complex leads to the catalytically reactive state. In some cases this can be due to a conformation change in the protein, in others a change in the substrate towards the transition change or a rearrangement of both partners (see discussion of 'on-enzyme equilibria' in Section 4.9). The history and detail of speculations and facts of the mechanisms have been reviewed by Jencks (1975, 1980). Ligand-induced conformation changes can also be the basis of mechanical work or the transmission of signals for some physiological response to messengers. As pointed out by Jencks, the potential affinity of the protein for the ligand must be greater than the measured association constant would indicate, since some of the binding energy is used for the conformation change of the protein. In the absence of ligand the conformational equilibrium

$$\text{P} \rightleftharpoons {}^{*}\text{P}$$

has a positive value of ΔG and in the presence of ligand

$$\text{PL} \rightleftharpoons {}^{*}\text{PL}$$

has a negative ΔG. If we describe the process by

$$\begin{array}{ccccc} & & {}^{*}\text{P} & & \\ & K_1 \nearrow & & \searrow K_2 & \\ \text{P} & & & & {}^{*}\text{PL} \\ & K_3 \searrow & & \nearrow K_4 & \\ & & \text{PL} & & \end{array}$$

then, as shown in Section 4.8, $K_1K_2 = K_3K_4$ and the Gibbs energy change of going from P to *PL is the sum of the Gibbs energies of the conformation change and the ligand binding process. The equilibrium constant of the overall process P $\rightleftharpoons$ *PL can be determined by titration or equilibrium dialysis, but the evaluation of the individual equilibrium constants requires kinetic analysis by observation of the amplitudes of the chemical relaxation of the system. This interesting approach to the study of binding equilibria is beyond the scope of the present volume, and the reader is referred to Winkler-Oswatitsch and Eigen (1979) for the treatment of relaxation amplitudes in terms of equilibrium constants and binding enthalpies. Results obtained by kinetic analysis of the conformation change alone over a range of ligand concentration can sometimes provide a good deal of information. For example in the case of horse liver alcohol dehydrogenase the kinetics of the interconversion of the enzyme–nucleotide complex is dependent on pre-equilibria with both nucleotide and proton binding. The nucleotide binds to the active site and this results in a decrease in the fluorescence of one of the two tryptophans of each of the two independently active subunits and a decrease in the pK of one, as yet unspecified, amino acid residue of each unit. Kinetic observations of these events reveal the following scheme in which only the different forms of the enzyme and its complexes are shown (Coates, Hardman, Shore, and Gutfreund, 1979):

$$\text{E–H} + \text{NAD}^+ \underset{K_1}{\overset{\text{I}}{\rightleftharpoons}} \text{E–H–NAD}^+ \underset{K_2}{\overset{\text{II}}{\rightleftharpoons}} {}^*\text{E–H–NAD}^+ \underset{K_3}{\overset{\text{III}}{\rightleftharpoons}} {}^*\text{E}^-\text{–NAD}^+ + \text{H}^+$$

(the starred forms designating the enzyme conformation with the quenched fluorescence). The equilibrium constants can be expressed at specified constant concentrations of NAD$^+$ and H$^+$:

$$K_1[\text{NAD}^+] = [\text{E–H–NAD}^+]/[\text{E–H}] = 1.47 \times 10^3[\text{NAD}^+];$$
$$K_2 = [{}^*\text{E–H–NAD}^+]/[\text{E–H–NAD}^+] = 2.52;$$
$$K_3/[\text{H}^+] = [{}^*\text{E}^-\text{–NAD}^+]/[{}^*\text{E–H–NAD}^+] = 2.51 \times 10^{-8}/[\text{H}^+].$$

In the absence of NAD$^+$ and at pH below neutrality the conformation change E–H $\rightleftharpoons$ *E–H is energetically unfavourable (i.e. results in a positive standard Gibbs energy change), while with NAD$^+$ bound to the active site the conformation change results in $\Delta G° = -2.3$ kJ. Clearly if NAD$^+$ binding favours the starred conformation, the starred conformation favours NAD$^+$ binding and the increased binding energy overcomes the unfavourable conformation change of the free enzyme. The whole process described above is, of course, an essential part of the energetics of the catalytic process. The binding energy contributes to the formation of a complex which will react with a lowered activation energy.

Some discussion of the contribution of the binding energy to the energy profile of catalytic reactions, as well as to processes coupled to the hydrolysis of ATP, as in transport and motility, was presented in Section 4.9.

5.7 A general treatment of ligand binding

In the previous sections the interpretation of ligand binding to identical and independent sites, and some of its consequences, were described phenomenologically. Some simple aspects of the general statistical treatment will now be discussed. This will serve to introduce examples for some concepts in statistical thermodynamics, like partition functions which are discussed in an elementary fashion in Section 2.5. Some of the results obtained in this section we have already derived earlier by a different route. Such repetitions can be helpful in giving a better understanding of the principles involved.

From definitions in the previous section it follows that for the reactions of ligand L binding to n sites on a molecule P, i.e.

$$P + L \rightleftharpoons PL_1, \quad \ldots \quad PL_{n-1} + L \rightleftharpoons PL_n,$$

the equilibrium constant for each of the reaction steps of binding the jth ligand molecule is

$$K_j = [PL_j]/[PL_{j-1}]\,[L]. \tag{5.12}$$

Since $[PL_j] = K_j[PL_{j-1}][L]$ and $[PL_{j-1}] = K_{j-1}\,[PL_{j-2}][L]$, it follows that

$$[PL_j] = K_1K_2 \ldots K_j[L]^j[P].$$

This equation is equivalent to the set of equations (5.11) if the intrinsic associations for the j steps are identical. From this we obtain Scatchard's expression for the concentration of the average number of ligand molecules bound per macromolecule P. For the case of non-interacting but not necessarily identical sites we define the total macromolecule concentration by

$$\begin{aligned}\sum_{\nu=0}^{n}[PL_\nu] &= [P] + [PL_1] + \ldots + [PL_n] \\ &= [P](1 + K_1[L] + K_1K_2[L]^2 + \ldots + K_1K_2 \ldots K_n[L]^n)\end{aligned}$$

and the total bound ligand concentration by

$$\begin{aligned}\sum_{\nu=0}^{n} \nu[PL_\nu] &= [PL_1] + 2[PL_2] + \ldots + n[PL_n] \\ &= [P]\sum_{\nu=0}^{n} \bar{K}_\nu\,[L]^\nu,\end{aligned}$$

where

$$\bar{K}_\nu = \prod_{j=1}^{\nu} K_j$$

and $K_o = 1$ by definition.

Thus for the average number of ligands associated per macromolecule we obtain

$$\begin{aligned}\bar{\nu} &= [P]\sum_{\nu=0}^{n} \nu\,\bar{K}_\nu\,[L]^\nu/[P]\sum_{\nu=0}^{n} \bar{K}\,[L]^\nu \\ &= \sum_{\nu=0}^{n} \nu\,\bar{K}_\nu\,[L]^\nu/\sum_{\nu=0}^{n} \bar{K}_\nu\,[L]^\nu.\end{aligned} \tag{5.13}$$

This equation should be compared with the expression for $\bar{\nu}$ (equation (5.1)) for a system with identical binding sites. In this general form it was derived by

Adair (1925a) for the case of oxygen binding by haemoglobin, where $n = 4$. He was the first to demonstrate, by osmotic pressure measurements, that the haemoglobin molecule was four times as large as most investigators had previously supposed, and contains four oxygen-binding (haem) groups.

The function

$$Q_{(n)} = 1 + \bar{K}_1[\mathrm{L}] + \bar{K}_2[\mathrm{L}]^2 + \ldots + \bar{K}_n[\mathrm{L}]^n = \sum_{\nu=0}^{n} \bar{K}_\nu[\mathrm{L}]^\nu \tag{5.14}$$

which occurs in the above equation is analogous to a partition function in statistical mechanics. The partition function is the sum of the probabilities of all possible states of a system relative to a reference state. The probability of any state with a specified degree of liganding is given by

$$\frac{[\mathrm{PL}_\nu]}{[\mathrm{P}]_{\mathrm{total}}} = \frac{\bar{K}_\nu[\mathrm{L}]^\nu}{Q_{(n)}}.$$

Here $K_o \equiv \bar{K}_o = 1$ by definition. Thus the first term in $Q_{(n)}$ gives the relative number of macromolecules with $\nu = 0$, normalized to unity. The next term, $\bar{K}_1[\mathrm{L}]$ gives the number for which $\nu = 1$, and so on for the higher terms.

It can be shown† that

$$\bar{\nu}_\mathrm{L} = \frac{\partial(\ln Q_{(n)})}{\partial(\ln [\mathrm{L}])} = \sum_{\nu=0}^{n} \nu\bar{K}_\nu[\mathrm{L}]^\nu / \sum_{\nu=0}^{n} \bar{K}_\nu[\mathrm{L}]^\nu. \tag{5.15}$$

Wyman (1965, 1967) has defined a function which we denote as the binding potential Λ.‡ This potential, in a system of n components, at constant temperature and pressure, is a function of the chemical potentials of all the components but one. That one component is taken as the reference component, and the amounts of the other components are specified relative to

†The derivation of equation (5.15) is as follows:

$$\begin{aligned}\frac{\mathrm{d}(\ln Q_{(n)})}{\mathrm{d}\,(\ln[\mathrm{L}])} &= \frac{1}{Q_{(n)}}\frac{\mathrm{d}\,Q_{(n)}}{\mathrm{d}\,(\ln[\mathrm{L}])} \\ &= \frac{[\mathrm{L}]\mathrm{d}Q_{(n)}}{\mathrm{d}[\mathrm{L}]Q_{(n)}} \\ &= \frac{1}{Q_{(n)}}\{\sum_{\nu=0}^{n} \nu\bar{K}_\nu[\mathrm{L}]^{\nu-1}\}[\mathrm{L}] \\ &= \frac{1}{Q_{(n)}}\sum_{\nu=0}^{n} \nu\bar{K}_\nu[\mathrm{L}]^\nu \\ &= \sum_{\nu=0}^{n} \nu\bar{K}_\nu[\mathrm{L}]^\nu / \sum_{\nu=0}^{n} \bar{K}_\nu[\mathrm{L}]^\nu = \bar{\nu}_\mathrm{L}\end{aligned}$$

‡Wyman originally used a capital Russian L (i.e. Л) as the symbol for the binding potential. We prefer to use only symbols from the Greek and English alphabets for symbols, so we have employed a capital Greek lambda (Λ) as the equivalent of Wyman's symbol.

it. Then the value of $\bar{v}$ for any other component (L) is given by differentiating the binding potential with respect to the chemical potential of that component:

$$v_L = \left(\frac{\partial \Lambda}{\partial \mu_L}\right)_{p,T,\mu_K} = \frac{\partial \Lambda}{RT\partial(\ln [L])} . \tag{5.16}$$

Here we have taken the concentration of L as equal to its activity, so that the increment $d\mu_L$ in its chemical potential is simply RT d(ln[L]). By comparing equations (5.15) and (5.16) it is obvious that the binding potential Λ, in a system of the type which we have been discussing, is simply equal to $RT \ln Q_{(n)}$, and Wyman refers to $Q_{(n)}$ as a 'binding polynomial'. We return to the use of these functions later in this chapter.

It is our purpose to use algebraic derivations only when they emphasize some physical phenomenon or to analyse experimental data. It is, for instance, important to stress that one of the concepts to be learned from partition functions and the probabilities of individual states is the relation between time and number averages (see the ergodic hypothesis in Section 2.5). Let us consider the difference between these two averages. One can think in terms of each binding site having a ligand bound for a certain fraction of the time: this fraction being equivalent to the probability given by equations (5.15) and (5.16). One can also think in terms of the state of the system at a given moment, counting the instantaneous value of v for each molecule at that moment, and averaging over all the molecules to get $\bar{v}$, or $y = \bar{v}/n$, for the whole system. If the number of molecules in the system is very large, the value of y, averaged over all the molecules at any instant, should be equal to the y value for any one molecule, averaged over a very long time. For further definitions of partition functions, ensembles, and other fundamental concepts of statistical mechanics the reader is referred to text-books such as those by Moore (1963) and Atkins (1978, Chapters 20 and 21).

Under strictly equilibrium conditions, for the large ensembles of binding systems normally considered in practice, the time and number average have equivalent consequences. However, as an aside, the following problem with important consequences in biological reactions should be mentioned. Suppose that a proton has to be bound to a particular basic group on an enzyme to make it catalytically active, and the hydrogen ion concentration of the medium corresponds to pH $=$ pK. If one were to take the number average and think of half the enzyme in the active form, this results in different (and erroneous) kinetic predictions. The correct interpretation for kinetic consequences is that every enzyme molecule is active a fraction of the time.

5.8 Systems with interaction between binding sites, and with sites of different intrinsic binding affinities

In the majority of actual situations, the binding sites on biological macromolecules may not be equivalent and independent. Quite commonly there are several distinct classes of such sites, with different association constants for a given ligand. In other cases, such as the successive binding of four oxygen or carbon monoxide molecules by haemoglobin, the binding sites may (as a good approximation) be considered equivalent, but there are strong interactions, so that binding at one site either increases or decreases the affinity of other sites for the ligand. In the former case, which is characteristic of haemoglobin, we call the binding cooperative; in the latter, which is characteristic of the binding of coenzymes by certain enzymes, we call it anticooperative. These terms require careful and critical discussion; we return to them below.

5.8.1 Sites of different intrinsic affinities: acid–base equilibria in proteins

Proton-binding equilibria in protein and nucleic acid molecules furnish an important example of classes of binding sites that differ greatly in proton affinity. The side chains of amino acid residues in proteins commonly contain seven kinds of ionizable groups that can bind or release protons. These are the carboxyl groups of aspartic and glutamic acids, the ε-amino group of lysine, the imidazole group of histidine, the guanidinium group of arginine, the sulphydryl group of cysteine, and the hydroxyl group of tyrosine. In addition there are the free α-amino groups at the NH_2 terminus of peptide chains, and the free α-carboxyls at the COOH terminus, or nine classes of groups in all. Table 5.1 lists characteristic pK values for each of these classes of groups, as they occur in simple peptides or other small molecules.

Proteins commonly contain many groups of some of these classes. Thus bovine serum albumin (molecular mass close to 65 000 daltons) contains about 100 aspartic and glutamic acid groups, 16 imidazole groups, 22 guanidinium groups, 19 phenolic tyrosyl groups, 57 ε-amino groups of lysine, one α-amino, one α-carboxyl, and one sulphydryl group. In a rough calculation we may take all the groups of a given class as having a single pK value, as if they were equivalent and independent, but this is only a preliminary estimate, that requires correction for at least two reasons:

(1) Addition of a proton to a protein, at a given pH, increases its net positive charge (or decreases its net negative charge) by one proton unit. This makes the net charge on the molecule more positive, and more electrical work must be done to bring up the next proton to another binding site. Thus the binding sites cannot be considered as really independent; there are negative interactions between them, due to electrostatic forces, which makes the curve for $\bar{\nu}_H$ as a function of pH less steep than it would be in the absence of such forces.

(2) The specific locations of particular groups in the folded protein structure may alter their pK values considerably. In hen egg lysozyme, for example, there is a particular carboxyl group with a pK value near 6, well above the normal range for free carboxyl groups. Some potential ionizing groups may be buried in the interior of the native protein, and will remain unreactive until the native structure becomes unfolded at low or high pH. This is true of a

considerable number (not all) of the imidazole groups in myoglobin and haemoglobin; they are not protonated until unfolding begins, around pH 4 and below. Likewise many tyrosyl hydroxyl groups in ovalbumin and other proteins remain unionized until the protein structure begins to unfold. For ovalbumin this does not happen until around pH 12 (see also discussion in Section 5.3).

Nevertheless the p*K* values in Table 5.1 furnish a valuable rough guide to the state of electric charge of a protein at physiological pH. We may expect, at pH 7, that nearly all the carboxyl groups will be deprotonated, and negatively charged. The guanidinium groups of arginine will almost invariably be positively charged, since their p*K* is extremely high. The ε-amino groups of lysine will in general be positively charged also, and the tyrosyl hydroxyls and cysteine sulphydryls uncharged. The imidazole groups of histidine have intrinsic p*K* values near 7, and are thus likely to be partly charged and partly uncharged at neutral pH. They generally furnish most of the physiological buffering action that proteins supply. The terminal α-amino groups of peptide chains of proteins also commonly have p*K* values between 7 and 8. In haemoglobin, for example, these groups play a significant part, both in buffer action and in the physiologically important CO_2 transport, by binding CO_2 in the form of carbamate.

In any case the different classes of basic groups in proteins differ enormously in proton affinity. The negatively charged ion of a tyrosyl group (p*K* near 10) binds protons about a million times as strongly as a carboxylate anion, with p*K* near 4. Thus the titration curve tends to be separated into regions of high buffering power, where groups of a given class are titrating, and intervening regions where there is little buffering.

These brief remarks are intended only as an elementary introduction to the field. For discussions in depth, see Edsall and Wyman (1958), Tanford (1961, 1962), and some recent papers by Gurd and his associates (Matthew, Hanania and Gurd, 1979; Friend and Gurd, 1979).

5.9 Interactions among equivalent (and nearly equivalent) sets of binding sites; cooperative and anticooperative interactions

We turn now to cooperative and anticooperative interactions, and will focus on systems containing a defined number of equivalent, or nearly equivalent, binding sites. These exemplify the important relations in their simplest form; they are also of central importance in biochemical systems. We consider first systems that exhibit cooperative interactions.

The binding of oxygen by haemoglobin furnishes a classic example of cooperativity. Figure 5.9 provides a typical binding curve, with y as ordinate and $[O_2]$ as abscissa. It is immediately obvious that the binding curve is very different from a rectangular hyperbola. Near $y = 0$ the curve rises with a rather small positive slope; as the oxygen pressure (activity) increases, the curve turns more steeply upward, and finally bends over as saturation approaches, with an

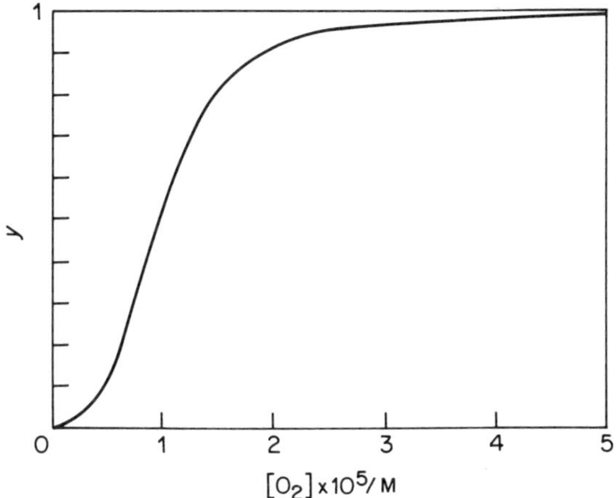

Figure 5.9 Simulated oxygen saturation curve for human haemoglobin using the macroscopic association constants of Ackers (1980): $K_1 = 4 \times 10^4$ M^{-1}; $K_2 = 2.73 \times 10^4$ M^{-1}; $K_3 = 5.90 \times 10^4$ M^{-1}; $K_4 = 202 \times 10^4$ M^{-1}

asymptotic approach to the limiting value of $y = 1$. The fact that the curve becomes increasingly steep as the ligand activity increases from zero is the essential feature in this plot that marks the curve as cooperative. Such a curve could not be given by a molecule such as myoglobin, for instance, which has only one binding site of exactly the same sort found in haemoglobin. As we know, the myoglobin curve must have the form of a rectangular hyperbola, and this means that the curve is steepest at the origin, and the slope steadily decreases as p_{O_2} increases. A curve like that for haemoglobin can arise only if there are at least two ligand binding sites. In this case we know from abundant evidence that there are four.

The significant fact about positive cooperativity is the sharp response to a small change in ligand concentration. We contrast this with the behaviour of a system of equivalent and independent sites, without interaction. In such a system the increase in ligand concentration required to go from $y = 0.1$ to $y = 0.9$ is given by

$$y = K[\mathrm{L}]/(1 + K[\mathrm{L}]).$$

For $y = 0.1$, $[\mathrm{L}] = 1/9K$; for $y = 0.9$, $[\mathrm{L}] = 9/K$. Thus there is an 81-fold change in [L] on going from $y = 0.1$ to $y = 0.9$. In the case of human haemoglobin we see from Figure 5.9 that the O_2 concentration needs to change only by a factor of about 3.5 to produce the same change in y. This is of immense physiological importance; it means that, as arterial blood passes into the capillaries, it will release a large amount of oxygen to the tissues in response to a relatively small drop in oxygen pressure; the cooperativity greatly increases the efficiency of unloading of oxygen in the tissues where it is needed. It is also important, of course, that the unloading must occur over a range of oxygen pressures that corresponds to the general conditions prevailing in the

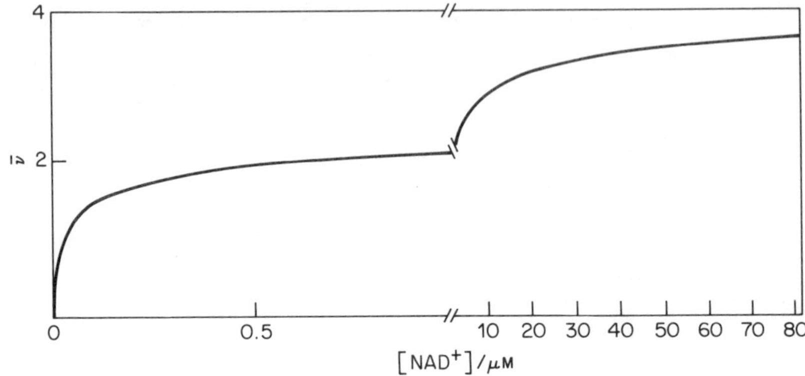

Figure 5.10 Titration curve for the binding of NAD^+ to rabbit muscle glyceraldehyde 3-phosphate dehydrogenase simulated from the association constants obtained by Bell and Dalziel (1975): 100×10^6 M^{-1}; 11.1×10^6 M^{-1}; 0.25×10^6 M^{-1}; 0.026×10^6 M^{-1}.

tissues. In blood the heterotropic interactions, discussed in Section 5.15, between oxygen binding and the binding of protons, CO_2, and organic phosphates in the red cell, make fundamental contributions to the efficient respiratory transport of oxygen and CO_2. These considerations are developed in much greater detail by Newsholme and Start (1973) and by Garby and Meldon (1977).

Clearly such cooperative interactions can make contributions to the sensitive response of receptors involved in many biological systems. Similar sharp transitions between two (or sometimes more) states of systems are discussed in Chapter 6.

There are also important examples in which binding to a system of initially equivalent sites is anticooperative: when one site is occupied, its neighbours become less receptive to the ligand. A well documented example is the binding of NAD^+ to glyceraldehyde 3-phosphate dehydrogenase, studied by Bell and Dalziel (1975). This enzyme, with four equivalent binding sites, one in each of four subunits, binds the first NAD^+ molecule with an association constant nearly four orders of magnitude greater than the fourth NAD^+. The constants obtained by Bell and Dalziel are used as an example in Figure 5.10; values obtained for the association of NAD^+ to rabbit muscle glyceraldehyde 3-phosphate dehydrogenase by other authors are quoted by Henis and Levitzky (1980). The binding curve derived from the constants of Bell and Dalziel shown in Figure 5.10 rises very steeply from $\bar{\nu} = 0$, and then flattens out, finally approaching saturation extremely slowly at high values of $[NAD^+]$. As Bell and Dalziel point out, the results could be interpreted either in terms of anticooperative interactions between initially identical sites or in terms of non-equivalent sites, the strongest binding sites of course reacting first with the ligand. The saturation curve cannot distinguish between the two forms of ligand binding. Chemically the four subunits of the enzyme are identical, but it is possible that they are functionally non-identical. The situation here is not altogether clear. Scatchard plots have the same shape (see p. 162) for systems with negative cooperativity as for those with independent but non-equivalent sites.

5.10 Choice of units of concentration in binding of gases by haemoglobin: temperature effects

The activity of oxygen, in such studies as these, can be expressed either as $[O_2]$, the concentration of free oxygen in the liquid phase, or as p_{O_2}, the partial pressure of oxygen in equilibrium with the solution. At any given temperature, these two quantities are proportional to each other by Henry's law (equation (3.66)) which can be taken as valid for the range of partial pressures we are considering – say below 1 atm. Thus the ratios of the successive binding constants (K values) will be the same on either basis, but the numerical values depend on the units chosen for pressure or concentration, and on the solubility coefficient of oxygen in the liquid phase. Values for oxygen in water, at temperatures from 0 to 50 °C, are given in Table 3.7; the value at 50 °C is almost exactly half that at 0 °C. The same is true for other gases, such as CO_2, that bind to haemoglobin. Also the dimensions of the binding constant K for the reaction $Hb + O_2 \rightarrow HbO_2$ are C^{-1} if we use concentration units, and p^{-1} if we use partial pressure units.

When we consider studies at different temperatures, we must distinguish clearly which units we are employing for the activity (or concentration) of the gas, since the Henry's law coefficient changes with temperature. If we are concerned with the enthalpy of combination in the reaction

(1) O_2 (gas) $\rightarrow$ O_2 (in solution), $\Delta H = -12\ \text{kJ mol}^{-1}$,

(2) O_2 (in solution) + Hb $\rightarrow$ HbO_2, $\Delta H = -59\ \text{kJ mol}^{-1}$.

The value of ΔH for reaction (1) (at 25 °C) is taken from Table 3.8. The value for reaction (2) is from Ackers (1980) as a rough mean of several values.

The enthalpy of reaction can be determined from the temperature coefficient of the binding constants by the van't Hoff equation (4.24): $d(\ln K)/dT = \Delta H/RT^2$ However, if the units of K are $(p_{O_2})^{-1}$ we are actually measuring the enthalpy of reaction for the sum of reactions (1) and (2): $\Delta H = -71\ \text{kJ mol}^{-1}$. If we express K in units of $(C_{O_2})^{-1}$, we obtain ΔH for reaction (2) only. Usually it is the latter quantity that primarily concerns us; in any case it is important to be clear as to which quantity we are actually calculating from the measurements.

5.11 Homotropic and heterotropic interactions

The curve in Figure 5.9 for oxygen binding of haemoglobin is an example of what Wyman terms homotropic interactions; the binding of one molecule of a given species modifies the binding affinity for other molecules of the same species. However, such binding curves can be greatly altered by changes in the concentration of other molecules or ions. Wyman denotes these as heterotropic interactions. Actually the first demonstration of homotropic interactions in oxygen binding, by Bohr *et al.* (1904), also contained a most striking effect of a heterotropic interaction. The sigmoid curves for oxygen binding, while keeping their general shape, moved progressively farther to the right with increasing partial pressure of CO_2. That is, added CO_2 diminishes the uptake of oxygen at a given p_{O_2} value. Bohr *et al.* immediately recognized the great biological significance of this relation for the transport of O_2 and CO_2 in the blood. Influx of CO_2 from the tissues into the capillary blood will decrease the oxygen-carrying capacity of the blood, and release large added amounts of oxygen to the tissues. In the lungs CO_2 is discharged from the blood, so p_{CO_2} falls, and this thereby increases the capacity of the blood to take up oxygen.

It was a logical inference from these observations that, if CO_2 affects the binding of O_2, then O_2 must affect the uptake of CO_2; thermodynamically the relation between the two effects is given by equation (3.10) and other relations that we shall give later. However, at the time neither Bohr nor any other investigator perceived this necessary relation, until it was established experimentally by Christiansen *et al.* (1914), 10 years after Bohr.

Shortly we shall consider several heterotropic interactions involving haemoglobin and its ligands – with protons, with the direct chemical binding of CO_2, and with organic phosphates such as diphosphoglycerate.

These heterotropic effects are to be distinguished from direct competitive effects between ligands that bind to the same site; the competition between the binding of O_2 and CO at the haem group is the best known example. The term 'heterotropic interactions' generally refers to the interactive effects of ligands that bind at different sites in the macromolecule, but in which the binding of either one of the different ligands affects the binding of the other. These are allosteric effects in the sense employed by Jacques Monod.

We turn first to the analysis of ligand-binding curves – the homotropic effects – and consider several techniques for evaluating the magnitude and character of cooperative and anticooperative interactions.

The following analysis applies specifically to systems with n equivalent binding sites, with interactions between them, where n is a known number. It is not suitable for systems such as serum albumin, in its interactions with anions, for which an important part of the analysis is the determination of n, for one or more classes of binding sites, as well as the binding constants involved. We have already discussed the use of the Scatchard plot in dealing with such systems. The Scatchard plot is also useful in analysing the behaviour of cooperative systems, and we give examples of its use below. First, however, we consider some other ways of analysing the binding curves of cooperative and anticooperative systems.

5.12 Deviations of binding groups from independent behaviour: the V (or $K_{\bar{x}}$) function

In a cooperative system the binding of ligand increases, with added ligand, more rapidly than for equivalent and independent sites; in an anticooperative system binding is depressed below that value. There are systems that are cooperative in one region of the binding curve, anticooperative in another. It is desirable, in analysing data, to plot them in a way that exhibits simply and clearly the kind of deviation from independent behaviour that actually exists. We treat first what is probably the most direct way of doing this, and then consider other approaches that are widely used.

First we note that, in a completely cooperative system, we may expect the successive intrinsic binding constants to increase, since binding at one site increases the affinity of others. Haemoglobin, and the other systems that we use for illustrations, have four binding sites. The Adair equation for $n = 4$ and ligand L becomes

$$\bar{\nu} = \frac{K_1[\mathrm{L}] + 2K_1K_2[\mathrm{L}]^2 + 3K_1K_2K_3[\mathrm{L}]^3 + 4K_1K_2K_3K_4[\mathrm{L}]^4}{1 + K_1[\mathrm{L}] + K_1K_2[\mathrm{L}]^2 + K_1K_2K_3[\mathrm{L}]^3 + K_1K_2K_3K_4[\mathrm{L}]^4}. \tag{5.17}$$

The generalization for n binding sites is obvious.

These K values are macroscopic binding constants, not the intrinsic constants for individual binding sites. The relation of the two kinds of constants is just the same as it is for equivalent and independent sites (equation (5.10)) except that we are now expressing relations in terms of association, rather than dissociation constants. For example the macroscopic constant K_1 is four times as large as the intrinsic constant k_1, since unliganded deoxyhaemoglobin has four unoccupied sites available for binding oxygen, whereas monoliganded haemoglobin has only one bound O_2 to release. Pursuing this reasoning we obtain the relations $k_1 = K_1/4$, $k_2 = 2K_2/3$, $k_3 = 3K_3/2$, $k_4 = 4K_4$.

We can rewrite the Adair equation in terms of the intrinsic constants. It becomes

$$\bar{\nu} = \frac{4(k_1[\mathrm{L}] + 3k_1k_2[\mathrm{L}]^2 + 3k_1k_2k_3[\mathrm{L}]^3 + k_1k_2k_3k_4[\mathrm{L}]^4)}{1 + 4k_1[\mathrm{L}] + 6k_1k_2[\mathrm{L}]^2 + 4k_1k_2k_3[\mathrm{L}]^3 + k_1k_2k_3k_4[\mathrm{L}]^4}. \tag{5.18}$$

Now we can define a function which, for a set of equivalent and independent groups, is a simple intrinsic constant, independent of $\bar{\nu}$. We denote it by the symbol V:†

$$V = \frac{\bar{\nu}}{(4 - \bar{\nu})[\mathrm{L}]} = \frac{k_1(1 + 3k_2[\mathrm{L}] + 3k_2k_3[\mathrm{L}]^2 + k_2k_3k_4[\mathrm{L}]^3)}{1 + 3k_1[\mathrm{L}] + 3k_1k_2[\mathrm{L}]^2 + k_1k_2k_3[\mathrm{L}]^3}. \tag{5.19}$$

It is obvious that, as [L] approaches zero, we obtain

$$\lim_{[\mathrm{L}] \to 0} V = k_1 = K_1/4,$$

†Earlier workers have denoted this quantity as Q, but we have reserved Q for the partition function of equation (5.14), and use V instead here.

and as [L] becomes very large, so that only the terms in $[L]^3$ are important, we have

$$\lim_{[L]\to\infty} V = k_4 = 4K_4.$$

A plot of V as a function of $\bar{\nu}$ will have a positive slope where binding is cooperative. For equivalent and independent sites it is a horizontal straight line. Moreover, if the values of $\bar{\nu}$ are obtained with high accuracy at the two ends of the curve ($\bar{\nu} = 0$ and $\bar{\nu} = 4$) it permits a preliminary estimation of k_1 and k_4. We note that this makes great demands on the reliability of the experimental technique, in order to know the value of $\bar{\nu}$ reliably when it is very small; or, with $\bar{\nu}$ close to 4, when the value of $4 - \bar{\nu}$ is very small.

If we can estimate k_1 and k_4 as described above, we can make a preliminary estimate of k_2 and k_3 from the limiting slope of log V as a function of $\bar{\nu}$ at the two ends of the curve:

$$\left.\begin{aligned} \lim_{\bar{\nu}\to 0} \frac{d(\ln V)}{d\bar{\nu}} &= 3(k_2 - k_1)/4k_1, \\ \lim_{\bar{\nu}\to 4} \frac{d(\ln V)}{d\bar{\nu}} &= 3(k_4 - k_3)/4k_3. \end{aligned}\right\} \qquad (5.20)$$

These limiting slopes can seldom be determined with high accuracy, but with the intercepts they permit a preliminary estimate of all the four intrinsic constants. To obtain the best values of the k_i requires the use of all the binding data, and their analysis by statistical methods such as those described in the appendix.

It was Scatchard who proposed this approach, which might be called 'the second Scatchard plot'; but this name might lead to confusion, so we do not use it. It was first applied, at Scatchard's suggestion, to the studies of Edsall, Felsenfeld, Goodman, and Gurd (1954) and Nozaki, Gurd, Chen, and Edsall (1957) on the binding of imidazoles by Cu^{2+} and Zn^{2+} ions. Here the ligand-binding substances are simple ions, not macromolecules, but the same principles apply. All sites are initially equivalent, and the number of sites is 4, as in haemoglobin. Figure 5.11(a) and (b) is from the work of Nozaki *et al.* on the binding of 4(τ)-methylimidazole (MeIm) to the two ions. Since the reactions occur in aqueous solution the reactions (for Cu^{2+}) are actually

$$Cu^{2+}(4H_2O) + MeIm \rightleftharpoons Cu^{2+}(3H_2O, MeIm) + H_2O$$

and so forth for both Cu^{2+} and Zn^{2+}. However, we usually ignore the water and write the reactions as

$$Cu^{2+} + MeIm \rightleftharpoons (Cu\ MeIm)^{2+}$$

and so forth.

The data on which Figure 5.11 is based are not of high accuracy, but they show a striking difference between the two ions in the binding of MeIm. Clearly the binding by Zn^{2+} is cooperative; that by Cu^{2+} is anticooperative, though it binds the first molecule of MeIm nearly 100 times as strongly as Zn^{2+} does. Since the curve for Zn^{2+} ascends as $\bar{\nu}$ increases, and that

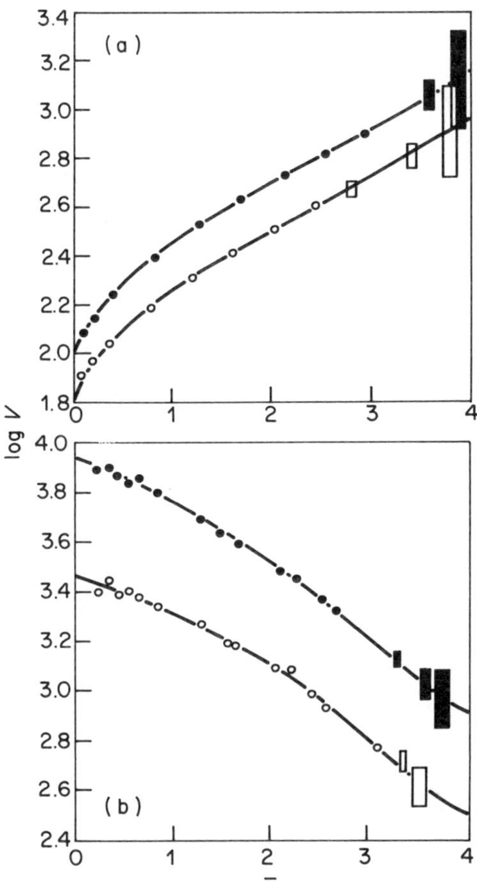

Figure 5.11 Plots of log V versus $\bar{\nu}$ for the binding of 4(τ)-methylimidazole to Zn^{2+} (a) and Cu^{2+} (b) at two different temperatures: ●, 2.5 °C; ○, 27 °C. (From Nozaki *et al.*, 1957.)

Table 5.2 Intrinsic binding constants for 4(τ)-methylimidazole at 25 °C

Binding ion	log k_1	log k_2	log k_3	log k_4
Cu^{2+}	3.53	3.31	3.05	2.56
Zn^{2+}	1.84	2.35	2.82	2.98

From Nozaki *et al.* (1957).

for Cu^{2+} descends, the fourth molecule of MeIm is bound with slightly greater strength by Zn^{2+}.

Nozaki *et al.* (1957) estimated, for the four intrinsic binding constants of Cu^{2+} and Zn^{2+} for 4(τ)-methylimidazole, the values given in Table 5.2. These show that for Cu^{2+}, $k_{i+1} < k_i$ for all successive values of i, whereas the opposite is true for Zn^{2+}.

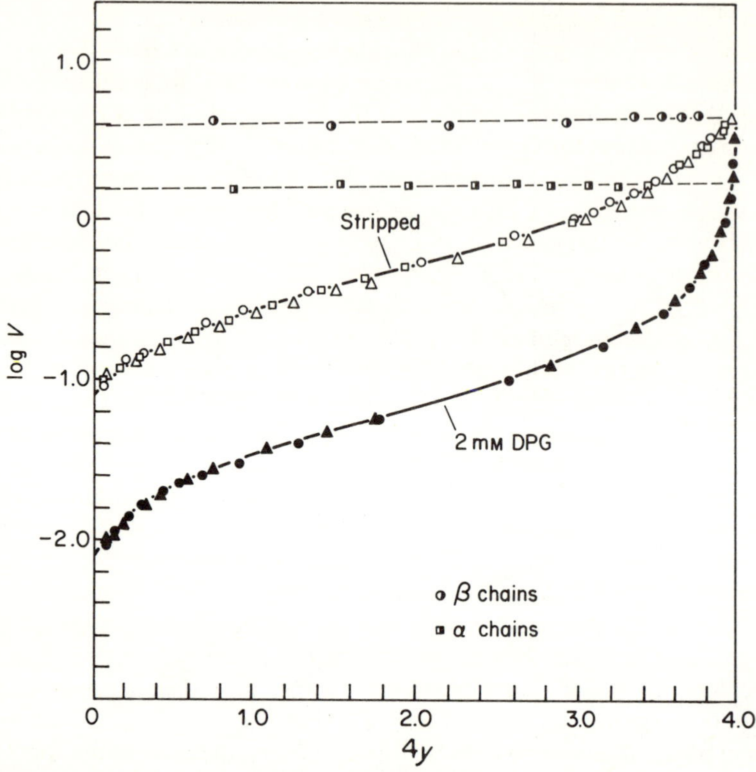

Figure 5.12 Plots of log V versus $4y$ for the oxygenation of stripped haemoglobin A in the presence and in the absence of diphosphoglycerate. The broken rules are for isolated chains in 0.1 M inorganic phosphate (pH 7.5, 20 °C) and are given for comparison. The solid rules are the simulated curves constructed by using the following values in terms of pressure of oxygen in millimetres of mercury:

	k_1	k_2	k_3	k_4	ΔG_{int}	n_{max}	p_{50}
Stripped HbA	0.079	0.295	0.75	4.35	9.6 kJ	2.52	1.9
2 mM DPG	0.008	0.037	0.02	4.35	15.4 kJ	3.02	15.3

These intrinsic association constants are not corrected for tetramer–dimer dissociation. This correction has been made in the work of Ackers described later in this chapter. (Adapted from Tyuma *et al.*, 1972.) For a definition of ΔG_{int} see p. 202.

In other words, binding to Cu^{2+} is *uniformly* anticooperative; to Zn^{2+} it is uniformly cooperative. In both cases the k_i increase or decrease by a factor of the order of 10 (1 in log k) in going from k_1 to k_4. Qualitatively similar sets of relations are found for these same ions in the binding of NH_3 and other nitrogenous ligands. For further data see for instance the table on p. 634 of Edsall and Wyman (1958) and the accompanying discussion.

We note that the four ligands to Zn^{2+} apparently bind in a tetrahedral arrangement; those that bind to Cu^{2+} form a square planar array.

The plot of log V as a function of $\bar{\nu}$ for human haemoglobin from the work of

Tyuma, Imai, and Schimizu (1972) is shown in Figure 5.12. The upper curve, for haemoglobin without organic phosphates, differs strikingly from the lower curve, for O_2 binding in the presence of 2×10^{-3} M diphosphoglycerate (DPG). It is immediately obvious that the binding of DPG depresses the first association constant for O_2 by a factor of about 10. Over most of the range of $\bar{\nu}$, the two curves run approximately parallel, but as $\bar{\nu}$ approaches 4 the curve in the presence of DPG rises steeply, and in the limit of 4 virtually coincides with the upper curve, indicating that k_4 is nearly the same for both. Independent binding measurements on DPG show that it binds rather strongly in a 1 : 1 ratio to the deoxyhaemoglobin tetramer, whereas its binding to oxyhaemoglobin ($\bar{\nu} = 4$) is very weak. This, like all heterotropic interactions, works both ways; DPG decreases the oxygen affinity, and oxygenation of haemoglobin drives off DPG. The form of the curves in Figure 5.12 would indicate that the oxygen affinity, in the presence of DPG, begins to rise rapidly from $\bar{\nu} = 3$ to $\bar{\nu} = 4$, which suggests that the release of DPG probably occurs chiefly in the last stage of oxygen binding. This is in accord with other evidence. It is well known that important conformation changes occur during the conversion of deoxy- to oxyhaemoglobin – see for instance Perutz (1980) and Baldwin and Chothia (1979) – and there is other evidence that the principal change occurs during the last stage of oxygen binding. The log V plot brings out this aspect of the oxygen binding curves more clearly than the direct plot of $\bar{\nu}$ against the concentration or partial pressure of oxygen.

This method of analysing data on ligand binding in terms of the function we have denoted as V can also be extended to data on the rates of enzyme-catalysed reactions, provided that we can assume that the reaction rates are a measure of a quasi-equilibrium between enzyme and substrate (see Section 5.1). Whitehead (1978) has given a comprehensive discussion of this method of analysing both categories of data. Actually Whitehead's function, which he denotes as $K_{\bar{x}}$, is the reciprocal of our function V, but the analysis is essentially the same. Whitehead gives a detailed analysis of the various methods that have been used for plotting such data, and shows how to calculate $K_{\bar{x}}$ (or its reciprocal, V) from these other plots. For further developments of the analysis, see also Whitehead (1980a,b).

We have noted, with respect to the intrinsic binding constants, that completely cooperative interactions are characterized by a sequence of intrinsic binding constants in ascending order: $k_1 < k_2 < k_3 \ldots < k_n$. For anticooperative interactions the reverse relation obviously holds. However, there are cases in which the k values do not increase or decrease in a uniform order; the binding may be cooperative for a certain range of values and anticooperative in others. Correspondingly the plots of V, or log V, against $\bar{\nu}$ or [L] may show maxima and minima, rather than being monotonic, as in Figures 5.11 and 5.12. Whitehead (1978) cites specific examples of such cases.

Finally we note again that this method of analysis of data requires that the number of binding sites, in the binding molecule or ion, be known. For such a system as the binding of chloride or other simple ions by serum albumin, n is not well defined. There are many weak binding sites, so weak that their actual numbers are uncertain, and a different approach has to be adopted.

The normal haemoglobin tetramer consists of two α and two β peptide chains, each with a haem group embedded in the tertiary structure. The two horizontal lines near the top of Figure 5.12 show the oxygen binding of the separated α and β chains. Plainly the interactions present in the $\alpha_2\beta_2$ tetramer have disappeared. For the α chains, which exist in solution as isolated chains, this is an obvious finding. The β chains, however, assemble into a tetramer, β_4, also known as haemoglobin H. However, even though they associate, they still function as separate oxygen-combining centres, as if they were four separate molecules; this is indeed an excellent example of a set of equivalent and independent binding sites. The great decrease in the first oxygen binding constant of the normal tetramer, $\alpha_2\beta_2$, as compared with the isolated chains, is immediately apparent from Figure 5.12. The deoxy conformation of $\alpha_2\beta_2$ is under strong internal constraints, which greatly diminish its affinity for oxygen; but these constraints diminish as binding proceeds, particularly when $\bar{\nu} = 3$ or above. The value of K_4 is comparable to that for the isolated chains; indeed recent work of Mills and Ackers (1979) indicates that it is slightly greater.

The normal tetramer of mammalian haemoglobins ($\alpha_2\beta_2$) undergoes reversible dissociation into dimers:

$$2\alpha\beta \rightleftharpoons \alpha_2\beta_2, \qquad K_{assoc} = [\alpha_2\beta_2]/[\alpha\beta]^2.$$

The value of K_{assoc} is much higher for deoxy- than for oxyhaemoglobin. The dimers lack cooperativity in binding, and their oxygen affinity is high, like that of the α monomer in Figure 5.12. (Further dissociation of $\alpha\beta$ into $\alpha + \beta$ is virtually negligible under all ordinary conditions.)

In order to obtain an accurate picture of the equilibrium of haemoglobin tetramers with oxygen we must correct for the presence of the dimers and their oxygen binding, which becomes increasingly important when haemoglobin is studied at high dilution. Ackers and his associates have studied this problem very thoroughly; the results are summarized by Ackers (1980), for human haemoglobin solutions of well defined composition (0.1 M Tris buffer pH 7.4, 0.1 M NaCl, 1 mM EDTA, 21.5 °C). The four intrinsic association constants, after correction for the presence of dimer, are $k_1 = 1 \times 10^4\ \text{M}^{-1}$, $k_2 = 1.82 \times 10^4\ \text{M}^{-1}$, $k_3 = 8.87 \times 10^4\ \text{M}^{-1}$, and $k_4 = 806 \times 10^4\ \text{M}^{-1}$, and the four macroscopic association constants are 4×10^4, 2.73×10^4, 5.91×10^4, and $201.5 \times 10^4\ \text{M}^{-1}$. The dramatic change comes when three sites are liganded, resulting in the intrinsic association constant for the fourth ligand being about two orders of magnitude greater than the other constants. This is in accord with the inferences we have drawn from the log V plots. Figure 5.9 shows a binding curve calculated from the Adair equation using the four macroscopic constants derived from Ackers' data as shown above. For a more detailed discussion of the linkage between ligand binding and macromolecular association, the reader is referred to Section 5.16.

It is illuminating to consider the relative concentration of the different species

Table 5.3 Distribution of species for ligand binding molecules with four equivalent sites when $\bar{v} = 2$ ($y = 0.5$)

Molecular species	Molecule with independent sites			Human haemoglobin		
	Relative conc., c	vc	v^2c	Relative conc., c	vc	v^2c
P ($v = 0$)	0.0625	0	0	0.370	0	0
PL ($v = 1$)	0.250	0.250	0.250	0.144	0.144	0.144
PL_2 ($v = 2$)	0.375	0.750	1.50	0.038	0.076	0.152
PL_3 ($v = 3$)	0.250	0.750	2.25	0.022	0.066	0.198
PL_4 ($v = 4$)	0.0625	0.250	1.00	0.426	1.704	6.816
	—	—	—	—	—	—
Sum of terms	1.000	2.00	5.00	1.000	1.99	7.31

See also Figure 5.13.
Haemoglobin distribution is calculated from binding constants given by Ackers (1980).
Note that the sum of the vc terms gives $\bar{v}$, since $c = 1.000$. Likewise the sum of the v^2c terms gives $\overline{v^2}$.
As shown later, the Hill coefficient n_H is given by

$$\frac{n}{\bar{v}(n - \bar{v})} = \frac{4}{\bar{v}(4 - \bar{v})} [\overline{v^2} - (\bar{v})^2]$$

Thus n_H for independent sites $= 5 - 4 = 1$. n_H for haemoglobin (at $\bar{v} = 2$) $= 7.31 - 4 = 3.3$.

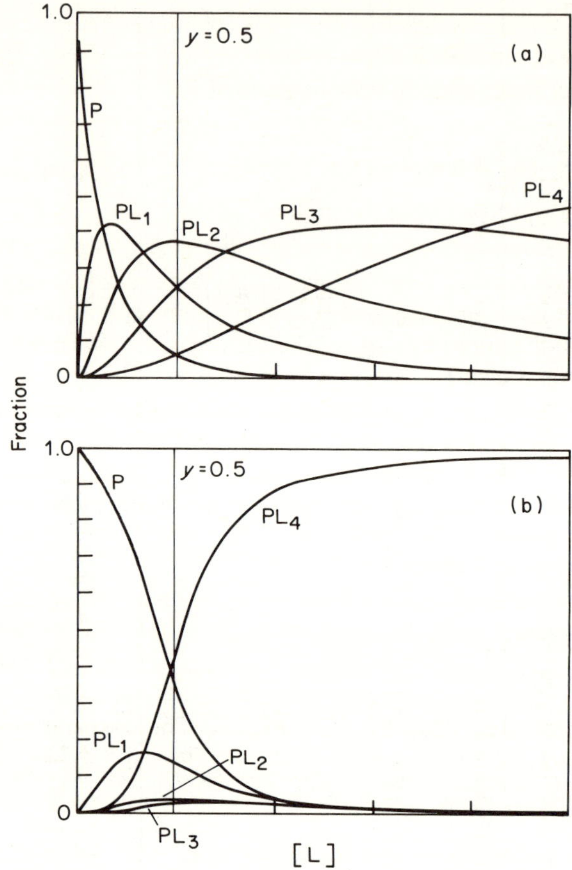

Figure 5.13 (a) Fractions of non-interacting intermediates for a system with identical K values: $K_1 = 4$, $K_2 = 3/2$, $K_3 = 2/3$, $K_4 = 1/4$. (b) Fraction of intermediates for binding of oxygen to haemoglobin using the K values from Figure 5.9

of unliganded and liganded haemoglobin, when on the average half the sites are occupied by oxygen ($y = 0.5$, $\bar{\nu} = 2$). There are five species with $\nu = 0, 1, 2, 3$, and 4, their relative amounts being given by the successive terms in the partition function Q (equation (5.14)), i.e. 1, $K_1[\mathrm{L}]$, $K_1K_2[\mathrm{L}]^2$, etc. If all sites were equivalent and independent, the relative distribution of the five species would be given by the binomial theorem, in the proportion 1 : 4 : 6 : 4 : 1, as we have already shown (p. 175). The actual distribution, for the two cases of independent sites and sites described by the Ackers constants, at $y = 0.5$, is shown in Table 5.3. In the latter case nearly 90% of the haemoglobin molecules, at $\bar{\nu} = 2$, are in the two extreme forms P ($\nu = 0$) and PL_4 ($\nu = 4$). Each of the intermediates PL_2 and PL_3 comprises less than 4% of the total distribution (Table 5.3); PL is present in slightly larger amounts. Table 5.3 shows the detailed calculation for the values of $\bar{\nu}$ and $\overline{\nu^2}$; the value of the latter is closely related to the Hill interaction coefficient, which we discuss in the next section (see also calculation of intermediate concentrations for systems with independent sites in Section 5.5).

5.13 The Hill plot and its relation to the Scatchard plot

A powerful method of analysis of binding curves that involve interacting sites arises from an equation originally proposed by A. V. Hill (1910, 1913) to interpret the behaviour of haemoglobin, at a time when this protein was believed to be a monomer, with only one oxygen binding site. Hill perceived that the observed sigmoid binding curve for oxygen could not be explained in terms of reaction at a single site, and he postulated the existence of reversible aggregation between such monomers, with strong interactions between binding sites in the aggregates, in order to explain the data. Later, Adair and Svedberg showed that haemoglobin is actually a tetramer. However, Wyman (1964, 1967) has developed the Hill equation, in a much more general way than that originally proposed by Hill, into a powerful tool for characterizing interactions in ligand binding.

Hill (1913) visualized ligand binding by haemoglobin as a completely cooperative process,

$$P + nL \rightleftharpoons PL_n,$$

with no intermediates PL_1, PL_2, ..., PL_{n-1}. The association constant for such a case is

$$K = [PL_n]/[P][L]^n.$$

Using $[P]_t$ for the total protein concentration, we obtain the logarithmic expression

$$\log K + n \log [L] = \log \{[PL_n]/([P]_t - [PL_n])\}.$$

Using n_H for the Hill constant and $y = [PL_n]/[P_t]$ we obtain

$$\log K + n_H \log [L] = \log (y/(1 - y))$$

for the commonly used logarithmic form of the Hill equation.

The Hill equation involves plotting the fraction y of bound ligand as a function of [L] in terms of the Hill coefficient n_H:

$$n_H = \frac{d \ln(y/(1 - y))}{d(\ln[L])} = \frac{1}{y(1 - y)} \frac{dy}{d(\ln [L])} . \tag{5.21}$$

If all sites were identical and independent, the resulting plot would be a straight line with $n_H = 1$. In the case of haemoglobin, with strong cooperative interctions among the binding sites, the value of n_H is generally relatively constant over a considerable range of values in the central range, say from around $y = 0.3$ to $y = 0.7$. In the range where such a simple relation holds, the data can be fitted by an equation with a single constant:

$$y = \frac{K[\mathrm{L}]^{n_\mathrm{H}}}{1 + K[\mathrm{L}]^{n_\mathrm{H}}} \quad \text{or} \quad \frac{y}{1-y} = K[\mathrm{L}]^{n_\mathrm{H}}. \tag{5.22}$$

This is Hill's original equation. Consider a hypothetical haemoglobin that could bind four oxygen molecules simultaneously, with no intermediates – only Hb ($\nu = 0$) and $Hb(O_2)_4$, for which $\nu = 4$, could exist. This would correspond to the reaction Hb + 4L $\rightleftharpoons$ HbL_4. The binding curve for such a molecule, with the maximum conceivable cooperativity, would be given by the above equation, with $n_H = 4$; i.e. n_H would equal n, the total number of binding sites. Actually, of course, no such simultaneous interaction of $4O_2$ molecules with a haemoglobin molecule could actually exist; there are always intermediate forms between $\nu = 0$ and $\nu = n$, although they may be present only in very small amounts in such a system as haemoglobin and oxygen at $\bar{\nu}$ near 2. In a vast number of studies on normal haemoglobins the value of n_H at $\bar{\nu} = 2$ has been found to lie in the range from about 2.5 to 3.3, showing immediately that there is a high degree of cooperative interaction. Some abnormal mutant haemoglobins have much lower n_H values; even a change in a single amino acid locus can drastically reduce the cooperative interactions. Haemoglobin H, composed of four β chains, gives $n_H = 1$; this molecule is still a tetramer, but the coupling that gives rise to cooperative interactions has vanished. In systems with anticooperative interactions, where the binding of one ligand makes the binding of others more difficult, the value of n_H becomes less than 1.

The value of n_H does not remain constant over the whole range of [L], except in the special case of equivalent and independent sites, where $n_H = 1$ throughout. In all cases, at the extremes of binding, when $\bar{\nu}$ approaches 0, or when it approaches the maximum value n (so that $n - \bar{\nu} = 0$), n_H approaches unity in the limit. This follows mathematically from the fact that, as [L] approaches zero, we enter a region in which only two species of macromolecule are present in significant amount: for haemoglobin these are Hb ($\nu = 0$) and HbO_2 ($\nu = 1$) and in this region only a single binding constant for a simple mass law equilibrium applies. Likewise, when $\bar{\nu}$ closely approaches n, the only species significantly present are $Hb(O_2)_3$ and $Hb(O_2)_4$, and again a simple mass law relation involving a single binding site governs the interaction (see Figure 5.14).

It may be helpful to illustrate the significance of n_H for the simple case of a molecule with only two binding sites, with association constants K_1 and K_2, for a ligand L. In this case the equation for y becomes

$$y = (K_1[\mathrm{L}] + 2\,K_1K_2[\mathrm{L}]^2)/2\,(1 + K_1[\mathrm{L}] + K_1K_2[\mathrm{L}]^2)$$

and thus

$$y/(1-y) = (K_1[\mathrm{L}] + 2\,K_1K_2[\mathrm{L}]^2)/(2 - K_1[\mathrm{L}]);$$

$$n_\mathrm{H} = \frac{\mathrm{d}(\ln y/(1-y))}{\mathrm{d}(\ln[\mathrm{L}])} = \frac{[\mathrm{L}](1-y)}{y}\,\frac{\mathrm{d}(y/(1-y))}{\mathrm{d}[\mathrm{L}]}.$$

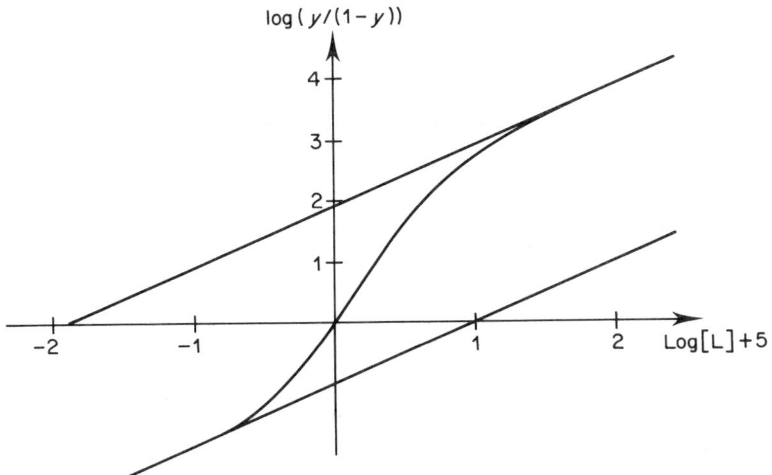

Figure 5.14 Simulation of the Hill plot for oxygen binding to human haemoglobin using the intrinsic association constants of Ackers (1980); see Figure 5.9. Slope = 3.33 at $y = 0.5$

The indicated differentiation yields the value

$$n_H = \frac{2(1 + 4K_2[L] + K_1K_2[L]^2)}{2 + (K_1 + 4K_2)[L] + 2K_1K_2[L]^2} . \tag{5.23}$$

We note that $n_H \to 1$ as $[L] \to 0$, and also as $[L] \to \infty$ (i.e. as $\bar{\nu} \to 2$), as we know it must, for any values of K_1 and K_2. In the special case of independent and equivalent sites, where $K_1 = 4K_2$, $n_H = 1$ for all values of [L]. At half saturation, when $y = 0.5$, the value of [L] at the midpoint of the titration, $[L] = [L_m] = (K_1K_2)^{-1/2}$. The value of n_H then becomes

$$n_{H,m} = 4/[2 + (K_1/K_2)^{1/2}] = 4/(2 + R).$$

Thus, when $R = (K_1/K_2)^{1/2} \to \infty$, $n_{H,m} \to 0$; when $R = 2$, $n_{H,m} = 1$ (equivalent sites); and when $R \to 0$, $n_{H,m} \to 2$.

It is often convenient to use a normalized ligand activity, $z = [L]/[L_m] = (K_1K_2)^{1/2}[L]$. Then the equation for n_H becomes

$$n_H = \frac{2(1 + 4R^{-1}z + z^2)}{2 + (R + 4R^{-1})z + 2z^2} . \tag{5.24}$$

There are actual systems that correspond closely to these limiting cases. For an example of very large R, consider a simple amino acid anion, such as glycinate, $H_2NCH_2COO^-$. Addition of one proton at 25 °C occurs with a binding constant of $10^{9.7}$ M^{-1} to form isoelectric glycine. The second proton binding constant is more than 10^7 times smaller, $10^{2.3}$; this yields the glycine cation. At the midpoint of the titration (the isoelectric point at pH 6.0) the binding curve as a function of pH is practically flat, and $n_H \simeq 0$.

The case in which $R = 2$ ($K_1 = 4K_2$) is exemplified approximately by a long chain α, ω-dicarboxylic acid, for which $n_H \simeq 1$ throughout the whole titration. The case in which $R \to 0$ is exemplified by many oxidation–reduction titrations; for instance, at high pH, where hydroquinone carries a high negative charge:

$$\text{quinone} + 2e^- \rightarrow \text{hydroquinone}^{2-}.$$

The two electrons appear to go on and off essentially simultaneously, with no detectable intermediate. This is true of many two-electron oxidation–reduction processes, although in many other cases the binding or release of the electrons occurs in two distinguishable steps, giving an intermediate free radical (a semiquinone). (See pages 107–111).

For the general case of n binding sites, Wyman (1964, p. 238 ff.) has defined what he terms the mean ligand activity (or concentration, if we take concentrations as equivalent to activities) $[L]_m$. In terms of the n successive binding constants, $K_1, K_2, \ldots, K_n$, Wyman (1967) has shown that

$$[L]_m = (K_1K_2 \ldots K_n)^{-1/n}. \tag{5.25}$$

When $n = 2$, this of course reduces to $(K_1K_2)^{-1/2}$, as we have already seen. In general $[L]_m$ corresponds fairly closely to the value of [L] when $y = 0.5$, but it can deviate somewhat from this when $n > 2$ if the successive binding constants are not symmetrical in their relations.

In any case it follows from the above equation and from the form of the binding polynomial $Q_{(n)}$ that when $[L] = [L]_m$ the concentration of the unbound macromolecule P must be equal to that of the completely saturated form PL_n.

The binding curve (Figure 5.9) provides values for y as a function of [L], the oxygen concentration, which have been used to simulate the Hill plot in Figure 5.14. The resulting curve has three characteristic slopes. The two asymptotes, at low and high values of y, correspond to binding to the first and fourth sites respectively. They have a slope of unity and extrapolate to log [L] $= -\log K$ when $\log (y/(1-y)) = 0$, thus giving values for k_1 and k_4, the intrinsic association constants. The maximum slope at $y = 0.5$ ($\bar{\nu} = 2$) is $n_H = 3.34$.

If really accurate data can be obtained at very low and very high degrees of saturation, the regions with unit slope become sufficiently distinct to provide enough information for the evaluation of the first and last association constants.

Before returning to a discussion of a further analysis of Hill plots in terms of the energy of interaction between sites, let us examine the relation between the Hill plot and the Scatchard plot. The Scatchard plot, for a system with cooperative interactions, has a positive slope at low [L] values. As [L] increases, the curve passes through a maximum and descends toward zero as [L] increases further. If one uses the data from the simulated haemoglobin saturation curve (Figure 5.9) for a Scatchard plot (Figure 5.15), one obtains a convex curve with a broad maximum around $y = 0.68-0.72$. Dahlquist

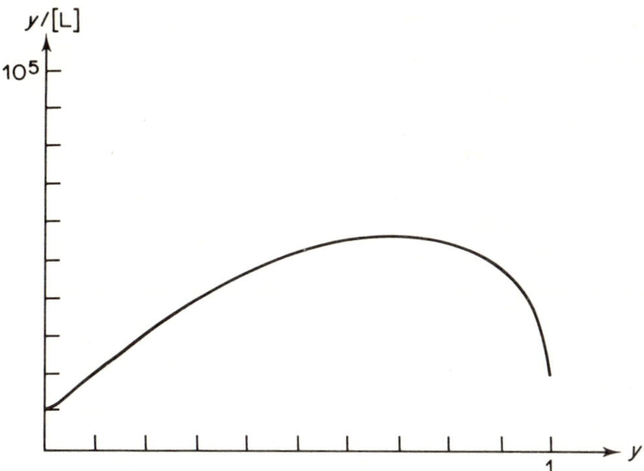

Figure 5.15 Simulated Scatchard plot for oxygen binding to human haemoglobin using the intrinsic association constants of Ackers (1980)

(1978) has shown the relation between the parameters of the Hill plot and the Scatchard plot. The slope of the Scatchard plot is

$$\frac{d(y/[L])}{dy} = \frac{1}{[L]}\left(1 - \frac{y}{[L]}\frac{d[L]}{dy}\right) = \frac{1}{[L]}\left(1 - \frac{d(\ln[L])}{d(\ln y)}\right). \tag{5.26}$$

The slope of the Hill plot is

$$\frac{d\ \ln(y/(1-y))}{d(\ln[L])} = \frac{[L]}{y}\left[\frac{1}{1-y}\right]\frac{dy}{d[L]} = \frac{1}{1-y}\left(\frac{d(\ln y)}{d(\ln[L])}\right).$$

If the Hill coefficient is evaluated at half saturation, where the curve is usually linear, one obtains

$$n_H = 4[L]\frac{dy}{d[L]} = 4\frac{dy}{d(\ln[L])}. \tag{5.27}$$

The maximum of the Scatchard plot is given by

$$\frac{d(\ln y)}{d(\ln[L])} = 1, \quad \text{for} \quad y = y_{max}.$$

The Hill slope at y_{max}, which is usually in the linear region near half saturation, is, from the general equation for the slope of the Hill plot,

$$n_H = (1 - y_{max})^{-1}. \tag{5.28}$$

Since the maximum (y_{max}) in the Scatchard plot is not sharply defined, but lies somewhere in the range 0.68–0.72, we can say only that n_H, as calculated in this way, must lie between 3.13 and 3.57. This should be compared with the value for n_H obtained directly from the slope of the Hill plot in the region near $y = 0.5$, which is 3.34. In this case the indirect calculation of n_H from the Scatchard plot is clearly less reliable than the direct calculation from the slope of the Hill plot; but it is of value to be aware of the relation between these two methods of analysing the data. Compare Figure 5.14 and 5.15 which are both constructed from the same data.

5.13.1 The Eadie plot for enzyme kinetics; the equivalent of the Scatchard plot

Seven years before Scatchard, Eadie (1942) had proposed plotting enzyme kinetic data by an equation mathematically identical with Scatchard's equation for reversible ligand binding. If $\bar{v}$ is the rate of the enzyme-catalysed reaction at substrate concentration [L], V_{max} the maximum rate at saturation with substrate, and K_m the Michaelis constant, then Eadie's equation can be written as follows:

$$\bar{v}/[L] = (V_{max} - \bar{v})/K_m = K_m^{-1}(V_{max} - \bar{v}).$$

It is K_m^{-1} that is the analogue to the binding constant K in Scatchard's equation, since K_m has the dimensions of a *dissociation* constant. Thus a plot of $\bar{v}/[L]$ against v is the analogue of a Scatchard plot in enzyme kinetics.†

The Eadie (Scatchard) plot has recently been shown by Hensley, Yang, and Schachman (1981) to be very useful in detecting small amounts of cooperative behaviour in enzymes. They were concerned particularly with the enzyme aspartate transcarbamylase (ATCase) of *Escherichia coli*, which catalyses the reaction

carbamyl phosphate + aspartate $\rightarrow$ carbamyl aspartate + phosphate.

This reaction is an essential early step in the biosynthesis of pyrimidines, and is under important heterotropic controls by cytidine triphosphate (CTP) and ATP. The former acts as an allosteric inhibitor, the latter as an activator.

In the presence of sufficient carbamyl phosphate, the reaction velocity, as a function of aspartate concentration, gives a sigmoid curve, qualitatively very similar to the oxygen-binding curve of haemoglobin. Structural analogues of aspartate, such as succinate, bind reversibly to ATCase, also with sigmoid binding curves. (For reviews of ATCase, see for instance Gerhart (1970) and Jacobson and Stark (1973).)

Some chemically modified derivatives of ATCase have greatly reduced cooperativity, and give velocity curves (or binding curves for substrate analogues) that are not easy to distinguish from hyperbolic curves for independently binding groups. Hensley *et al.* (1981) have shown that the Hill plot also gives little indication of cooperativity in such cases, but the Eadie

†Dixon and Webb (1964) have called attention to the fact that the Eadie plot was first proposed by Barnet Woolf in Cambridge, England, and was presented by Haldane and Stern (1932) – the German version of Haldane's well known book on enzymes (Haldane, 1930). It might therefore be called the Woolf plot, but we will continue to call it the Eadie plot, to avoid confusion.

(Scatchard) plot gives clear indication that cooperative interactions are involved. In exploring for systems showing cooperative interactions, therefore, this is likely to be a valuable approach.

5.13.2 Relation of the Hill coefficient to the distribution of binding

The Hill coefficient, n_H, is determined by the distribution of bound ligand among the macromolecules that bind it. We can show this by remembering the definition of n_H and differentiating the Adair equation:

$$\begin{aligned} n_H &= \mathrm{d}\{\ln[\bar{\nu}/(n-\bar{\nu})]\}/\mathrm{d}(\ln[\mathrm{L}]) \\ &= \frac{n}{\bar{\nu}(n-\bar{\nu})}\frac{\mathrm{d}\bar{\nu}}{\mathrm{d}(\ln[\mathrm{L}])} \\ &= \frac{n}{\bar{\nu}(n-\bar{\nu})}\frac{\mathrm{d}}{\mathrm{d}(\ln[\mathrm{L}])}\left(\Sigma i\bar{K}_i[\mathrm{L}]^i \sum_{i=0}^{n} \bar{K}_i[\mathrm{L}]^i\right) \\ &= \frac{n}{\bar{\nu}(n-\bar{\nu})}\left\{\frac{\Sigma i^2\bar{K}_i[\mathrm{L}]^i}{\Sigma\bar{K}_i[\mathrm{L}]^i} - \left(\frac{\Sigma i\bar{K}_i[\mathrm{L}]^i}{\Sigma\bar{K}_i[\mathrm{L}]^i}\right)^2\right\} \\ &= \frac{n}{\bar{\nu}(n-\bar{\nu})}\{\overline{\nu^2} - (\bar{\nu})^2\}. \end{aligned} \tag{5.29}$$

Now $\overline{\nu^2} - (\bar{\nu})^2$ is proportional to the standard deviation of the distribution of bound ligand. K. Linderstrøm-Lang was the first to derive this relation (for $\mathrm{d}\bar{\nu}/\mathrm{d}(\ln[\mathrm{H}^+]) = -(1/2.3)\mathrm{d}\bar{\nu}/\mathrm{dpH}$) for the proton binding curves of proteins and polyelectrolytes in general (see Cohn and Edsall, 1943, p. 462 and the general discussion on pp. 460–468). Equation (5.29) shows that a high n_H value is associated with a markedly non-random distribution of values among the binding macromolecules. The calculation of $\overline{\nu^2}$ and $(\bar{\nu})^2$ is illustrated in Table 5.3 for haemoglobin and for a molecule showing random binding of ligand, for the case $n = 4$, and $\bar{\nu} = n/2 = 2$. For haemoglobin the value of n_H is 3.3; for random binding it is unity, as indeed it must be for all $\bar{\nu}$ values of equivalent and independent binding sites, over the entire range of binding.

5.14 Some other approaches to analysis of ligand binding equilibria

Numerous approaches have been proposed to the analysis of ligand binding data. We cannot discuss all of them here, but we note a few significant references. Klotz and Hunston (1971) have analysed several methods of plotting data for the binding of small molecules by macromolecules with many classes of binding sites.

Gill, Gaud, Wyman, and Barisas (1978) have shown how various properties of the Hill plot, such as its asymptotes and derivatives, are expressible in terms of the mole fractions (x_i) of the macromolecule binding i molecules of ligand. Given sufficient information, the linear equations in the x_i can be solved to give the mole fractions of each of the various species: P, PL_1, PL_2, etc. The values of the four Adair constants can then be determined. Like any other method of evaluating the Adair constants, this method requires highly accurate data in the regions where $\bar{\nu}$ approaches zero or the saturation value ($\bar{\nu} = n$).†

Imai and Adair (1977) have described a relatively simple method for calculating the fractions of the various species of macromolecule with values of $\bar{\nu}$ from the experimental data, thence obtaining the Adair constants. This appears to give reliable values for binding constants by a very straightforward procedure.

5.14.1 Gibbs energy of interaction between binding sites

What Wyman (1964) calls the free energy (Gibbs energy) of interaction (ΔG_{I}) between the binding sites is defined by the relation

$$\Delta G_{\mathrm{I}} = RT \ln (k_n/k_1).$$

It is taken as positive when the interactions are stabilizing, i.e. when $k_n > k_1$. For Ackers' data on human haemoglobin under the conditions defined above (p. 192), $k_4/k_1 = 806$, which gives

$$\Delta G_{\mathrm{I}} = 16.59 \text{ kJ mol}^{-1} = 3.965 \text{ kcal mol}^{-1}.$$

The reader is referred to other transformations of the data suggested by Wyman (1964) and Whitehead (1978) for the calculation of interaction energies.

A valuable general treatment of ligand binding by macromolecules, from a somewhat different approach, is given by Weber (1975).

5.15 Linked functions and heterotropic interactions

We have discussed the cooperative and anticooperative interactions that can occur when a macromolecule binds successively several ligands of the same kind at different sites. Wyman has termed such interactions homotropic. There are also heterotropic interactions of great importance, when the binding of protons by haemoglobin alters its oxygen affinity, and conversely the oxygenation of haemoglobin alters its proton affinity (the Bohr effect). Such interactions are, in Wyman's term, linked, and thermodynamics requires that certain important reciprocal relations must hold between the binding of the different ligands.

As a simple example consider a macromolecule M which can bind two different ligands, A and B, each at a distinct binding site:

†Unfortunately there are several misprints in the equations in this article; the user of this method should check the derivations of the equations carefully.

$$\begin{array}{ccccc} & & MA & & \\ & {}^{K_A}\nearrow & & \searrow^{K_{AB}} & \\ M & & & & BMA \\ & {}_{K_B}\searrow & & \nearrow_{K_{BA}} & \\ & & BM & & \end{array}$$

Then

$$K_A = \frac{[MA]}{[M][A]}, \qquad K_B = \frac{[BM]}{[M][B]}$$

$$K_{BA} = \frac{[BMA]}{[BM][A]}, \qquad K_{AB} = \frac{[BMA]}{[B][MA]}$$

$$K_A K_{AB} = \frac{[BMA]}{[M][A][B]} = K_B K_{BA}.$$

If binding at the two sites is unlinked, then $K_A = K_{BA}$ and $K_B = K_{AB}$. However, if, for instance, K_A is found to be 20 times as great as K_{BA} – i.e. if the binding of B to M decreases the affinity of the A-binding site for A by a factor of 20 – then it follows from the equality relations between the K_i that K_B must also be 20 times as great as K_{AB}. Thermodynamics alone cannot tell us whether such a heterotropic interaction will be found in any particular case; but it does require that, if A influences the binding of B, then B must influence the binding of A; and it specifies the relation that must hold between them. If for instance M is haemoglobin, A is O_2, and B is H^+ ion, the above relations give a qualitative description of what happens when haemoglobin interacts with oxygen or acid in the physiological pH range. Haemoglobin, however, has four O_2 binding sites and several 'haem-linked' proton binding sites (most of the many acidic and basic groups in the molecule are not haem-linked), so we need a more elaborate analysis to describe it; this will be considered below. Myoglobin, which has only one O_2 binding site, shows no significant evidence of linkage between oxygen binding and the acid and basic groups.

We have already pointed out the significance of the polynomial $Q_{(n)}$ for the binding of a ligand L by a macromolecule (p. 179):

$$Q_{(n)L} = 1 + \bar{K}_{1L}[L] + \bar{K}_{2L}[L]^2 + \ldots + \bar{K}_{nL}[L]^n.$$

Here K_{iL}, as previously defined, denotes the product

$$\bar{K}_{iL} = K_{1L}K_{2L} \ldots K_{iL}$$

and $\bar{\nu}_L$ is given by the relation

$$\bar{\nu}_L = \partial(\ln Q_{(n)L})/\partial(\ln [L]) = (\partial\Lambda/\partial\mu)_{p,T,\mu_K}.$$

Here again Λ is the binding potential of Wyman, and $Q_{(n)L}$ is the binding polynomial for component L.

Consider now another ligand, Z, for which the macromolecule has (say) r binding sites. Then the binding polynomial for Z is

$$Q_{(r)Z} = 1 + \bar{K}_{1Z}[Z] + \bar{K}_{2Z}[Z]^2 + \ldots + \bar{K}_{rZ}[Z]^r. \qquad (5.30)$$

If both ligands are present, the macromolecule M can exist in $(n+1)(r+1)$ different liganded forms, from M_0 (unliganded) to ML_nZ_r, and the binding polynomial for M in all forms becomes

$$Q_{(n)L,(r)Z} = [M_0] \sum_{i=0}^{n} \sum_{j=0}^{r} \bar{K}_{ij} [L]^i [Z]^j.$$

For brevity we denote this function simply as Q. By the same reasoning as before, this gives

$$\bar{v}_L = \{\partial(\ln Q)/\partial(\ln [L])\}_{[Z]},$$
$$\bar{v}_Z = \{\partial(\ln Q)/\partial(\ln [Z])\}_{[L]}.$$

Since $d(\ln Q)$ is a perfect differential, as defined in Chapter 3, this immediately gives the relation

$$\left(\frac{\partial \bar{v}_L}{\partial(\ln[Z])}\right)_{[L]} = \left(\frac{\partial \bar{v}_Z}{\partial(\ln[L])}\right)_{[Z]} = \frac{\partial^2(\ln Q)}{\partial(\ln[L])\partial(\ln[Z])}. \qquad (5.31)$$

This is the basic equation of linkage. It can be readily extended to systems involving more than two ligands, as Wyman (1964) has described in detail. If for instance L is O_2 and Z is H^+ ion, equation (5.31) shows how the variation of $\bar{v}_{O_2}$ with pH at constant $[O_2]$ relates to the variation of bound H^+ ion with $\ln[O_2]$ at constant pH.

These linkage relations can be thrown into various useful forms, depending on which variables we wish to hold constant in a particular study. For a system involving two ligands, as above, we have four variables to consider – $\bar{v}_L$, $\bar{v}_Z$, $\ln[L]$, and $\ln[Z]$ – but only two of them can be varied independently. We can thus choose any two of the four as independent variables, and express the variation of the others in terms of them. For instance we may choose $\ln[L]$ and $\ln[Z]$ as the independent variables, and express variations in $\bar{v}_L$ and $\bar{v}_Z$ in terms of them. By the rules of partial differentiation (see Chapter 3) we have the relations

$$d\bar{v}_L = \left[\frac{\partial \bar{v}_L}{\partial(\ln[L])}\right]_{[Z]} d(\ln[L]) + \left[\frac{\partial \bar{v}_L}{\partial(\ln[Z])}\right]_{[L]} d(\ln[Z]), \qquad (5.32)$$

$$d\bar{v}_Z = \left[\frac{\partial \bar{v}_Z}{\partial(\ln[L])}\right]_{[Z]} d(\ln[L]) + \left[\frac{\partial \bar{v}_Z}{\partial(\ln[Z])}\right]_{[L]} d(\ln[Z]). \qquad (5.33)$$

By holding some of the variables constant, and varying others, we can obtain various useful relations that are directly applicable to experimental studies.

For instance, if we keep $\bar{v}_L$ constant ($d\bar{v}_L = 0$), we can get a relation between the activities of L and Z and the binding of these two ligands. Setting $d\bar{v}_L = 0$ in equation (5.32) and rearranging terms we obtain, by the procedure described in Chapter 3, the relation

$$\left[\frac{\partial(\ln[Z])}{\partial(\ln[L])}\right]_{\bar{v}_L} = -\left[\frac{\partial\bar{v}_L}{\partial(\ln[L])}\right]_{[Z]} \bigg/ \left[\frac{\partial\bar{v}_L}{\partial(\ln[Z])}\right]_{[L]} \tag{5.34}$$

$$= -\left[\frac{\partial\bar{v}_L}{\partial(\ln[L])}\right]_{[Z]} \bigg/ \left[\frac{\partial\bar{v}_Z}{\partial(\ln[L])}\right]_{[Z]} = -\left(\frac{\partial\bar{v}_L}{\partial\bar{v}_Z}\right)_{[Z]} .$$

The step indicated by the second equality sign leaves the numerator of the indicated ratio unchanged, but changes the denominator by substituting an equivalent expression from the fundamental linkage equation (5.31). Both the numerator and denominator of the resulting expression then refer to increments of $\bar{v}_L$ or $\bar{v}_Z$ with variation in the activity of L, with the activity of Z held constant. The final form of the equation then follows directly.

The same procedure, applied to equation (5.33) gives a similar equation:

$$\left[\frac{\partial(\ln[Z])}{\partial(\ln[L])}\right]_{\bar{v}_Z} = -\left(\frac{\partial\bar{v}_L}{\partial\bar{v}_Z}\right)_{[L]} . \tag{5.35}$$

Here the variables being held constant are $\bar{v}_Z$ on the left-hand side and [L] on the right; in the previous equation they were $\bar{v}_L$ and [Z] respectively.

Another fundamental linkage relation is

$$\left(\frac{\partial(\ln[L])}{\partial\bar{v}_Z}\right)_{\bar{v}_L} = \left(\frac{\partial(\ln[Z])}{\partial\bar{v}_L}\right)_{\bar{v}_Z} . \tag{5.36}$$

This follows from the previous fundamental linkage equation, if we interchange ln[L] with $\bar{v}_L$ and ln[Z] with $\bar{v}_Z$. The interchange is valid, since we could equally well have written the equations (5.32) and (5.33) in terms of $\bar{v}_L$ and $\bar{v}_Z$ as the independent variables, instead of ln[L] and ln[Z]. Likewise, from the symmetry of the equations, we can interchange $\bar{v}_L$ and $\bar{v}_Z$, and simultaneously interchange ln[L] and ln[Z], and the resulting equation will be valid (see Wyman, 1964, p. 227). It is important, in writing such equations, to take great care in noting just which variables are being held constant in any given case.

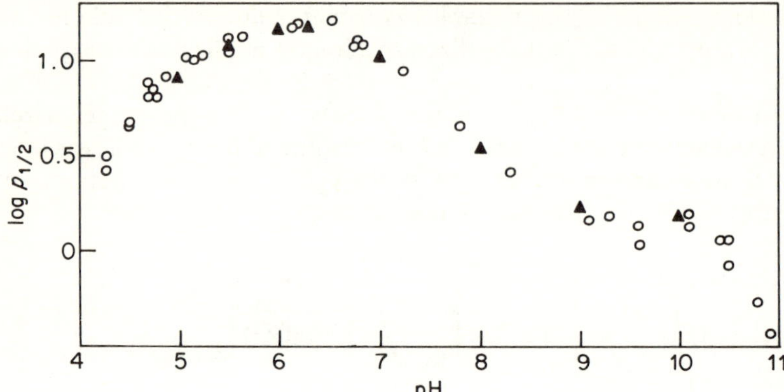

Figure 5.16 Comparison of directly observed (○) values of log $p_{1/2}$ with those calculated (▲) from differential titration. For discussion see Rossi Fanelli, Antonini, and Caputo (1964

These equations are directly applicable to the Bohr effect in haemoglobin, i.e. the release of protons on binding of oxygen at physiological pH, and the reciprocal influence of oxygen binding in decreasing the pK values of certain 'haem-linked' acid groups, thus causing a release of protons. We take [Z] as H^+ ion activity, with $\ln[Z] = -2.3$ pH, and [L] as oxygen activity, expressed as molar concentration in solution:

$$-\left[\frac{\partial(\ln[H^+])}{\partial(\ln[O_2])}\right]_{\bar{v}_{O_2}} = \left[\frac{2.3\ \partial pH}{\partial(\ln[O_2])}\right]_{\bar{v}_{O_2}} = \left(\frac{\partial \bar{v}_{O_2}}{\partial \bar{v}_{H^+}}\right)_{pH}, \tag{5.37}$$

and equation 5.36 becomes

$$\left(\frac{\partial(\ln[H^+])}{\partial \bar{v}_{O_2}}\right)_{v_{H^+}} = -2.3\left(\frac{\partial pH}{\partial \bar{v}_{O_2}}\right)_{\bar{v}_{H^+}} = \left(\frac{\partial(\ln[O_2])}{\partial \bar{v}_{H^+}}\right)_{\bar{v}_{O_2}}$$

The effect is measured experimentally in two different ways:

(1) Deoxyhaemoglobin is completely oxygenated, at a specified pH, and the solution is then titrated with acid or alkali until the original pH is restored. This titration determines how many protons are released (or bound) per oxygen molecule bound at that pH. This titration is repeated over a series of pH values, to map out how the effect varies with pH.

(2) The concentration of oxygen needed for half saturation of the haemoglobin ($\bar{v}_{O_2} = 2$) is determined over a range of pH values.

From the data in either one of these experiments we can infer what the result of the other will be. Figures 5.16 and 5.17 indicate the nature of the experimental results; the first of these figures shows the data for the variation of $\ln[O_2]$ with pH, when $\bar{v}_{O_2}$ is held constant at 2. (The corresponding partial pressure

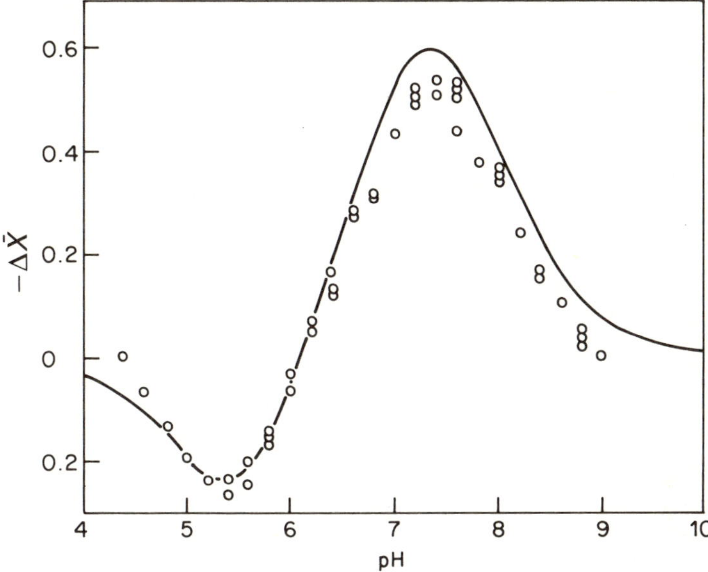

Figure 5.17 Differential proton uptake ($\Delta\overline{X}$) per haem group of oxygenated haemoglobin. (For details see German and Wyman, 1937.)

of oxygen, to maintain this half-saturation of haemoglobin, is commonly designated $p_{1/2}$. The concentration of O_2 in solution is, of course, by Henry's law, directly proportional to $p_{1/2}$.) It also shows the points calculated for this relation from the first type of experiment described above (the release or binding of H^+ ions on oxygenation at constant pH). The two types of experiment do indeed give equivalent results, as predicted.

We note that other linked relations exist in the haemoglobin system. There is an interaction at constant pH between CO_2 binding and oxygen binding, the CO_2 being bound as carbamate to the *N*-terminal amino groups of the α-chains, with decrease of the oxygen affinity. Correspondingly oxygenation of the haemoglobin diminishes the binding of CO_2 as carbamate. Likewise the binding of organic phosphates, notably 2,3-diphosphoglycerate, markedly decreases the oxygen affinity; and conversely the oxygenation of haemoglobin causes a release of most of the bound diphosphoglycerate. These experimental findings show the existence of other linked functions in the binding of ligands by haemoglobin, and all these processes are linked to one another. It is our aim here simply to introduce these relations, not to attempt a really searching study of them.

5.16 Linkage between macromolecular association and ligand binding

If a macromolecule, composed of subunits, undergoes reversible dissociation and association reactions, the equilibrium constant for such reactions may be altered by ligand binding. This must occur if the different forms of the

Tetramer ($\alpha_2\beta_2$) $\rightleftharpoons$ Two dimers ($2\alpha\beta$)
DeoxyHb ($\alpha_2\beta_2$) + 4L $\rightleftharpoons$ OxyHb ($\alpha_2\beta_2L_4$)
Deoxydimer ($\alpha\beta$) + 2L $\rightleftharpoons$ Oxydimer ($\alpha_2\beta_2L_2$)

Gibbs energy values ($\Delta G°$) in kilojoules per mole are as follows:

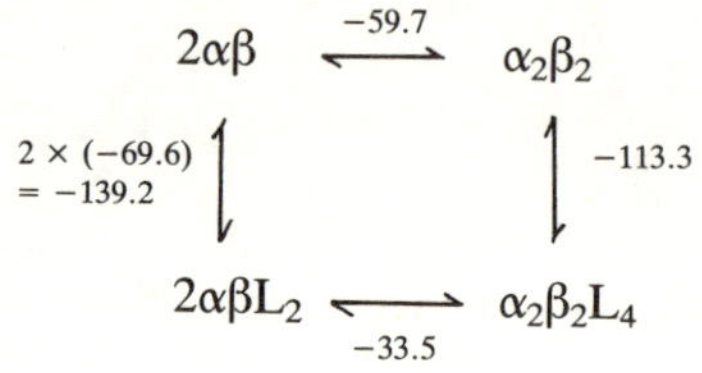

$$\begin{array}{ccc} 2\alpha\beta & \xleftrightarrow{-59.7} & \alpha_2\beta_2 \\ {\scriptstyle 2\times(-69.6) = -139.2}\updownarrow & & \updownarrow{\scriptstyle -113.3} \\ 2\alpha\beta L_2 & \xleftrightarrow[-33.5]{} & \alpha_2\beta_2L_4 \end{array}$$

Thus total Gibbs energy values are

$$\Delta G° = -59.7 - 113.3 = -173.0 \quad \text{(via } \alpha_2\beta_2\text{)},$$
$$\Delta G° = -139.2 - 33.5 = -172.7 \quad \text{(via } 2\alpha\beta L_2\text{)}.$$

Equilibrium constants $K_i = \exp\{-\Delta G_i°/RT\}$ may be calculated as follows:

$$[\alpha_2\beta_2]/[\alpha\beta]^2 = 3.9 \times 10^{10}\ \text{M}^{-1}, \quad [\alpha_2\beta_2L_4]/[\alpha_2\beta_2][L]^4 = 1.02 \times 10^{20}\ \text{M}^{-4},$$
$$[\alpha\beta L_2]/[\alpha\beta][L]^2 = 2.9 \times 10^{12}\ \text{M}^{-2}, \quad [\alpha_2\beta_2L_4]/[\alpha\beta L_2]^2 = 8.7 \times 10^{5}\ \text{M}^{-1}.$$

$$\left(\frac{[\alpha\beta L_2]}{[\alpha\beta][L]^2}\right)^2 \frac{[\alpha_2\beta_2L_4]}{[\alpha\beta L_2]^2} = \frac{[\alpha_2\beta_2L_4]}{[\alpha\beta]^2[L]^4} = \frac{[\alpha_2\beta_2]}{[\alpha\beta]^2}\,\frac{[\alpha_2\beta_2L_4]}{[\alpha_2\beta_2][L]^4}$$

$$= 4.17 \times 10^{30}\ \text{M}^{-5} \quad \text{(left-hand equation)}$$
$$= 3.98 \times 10^{30}\ \text{M}^{-5} \quad \text{(right-hand equation)}.$$

Figure 5.18 Linkage between tetramer–dimer equilibrium in human haemoglobin A and the binding of oxygen by the haemoglobin molecule. Data from Mills and Ackers (1979) and Ackers (1980). The conditions of the experiments were pH 7.4, 0.1 M Tris buffer, 0.1 M NaCl, 1 mM EDTA, at 21.5 °C. No phosphates were present. The diagram given by Ackers (1980, p. 333) gives further details involving the intermediates in ligand binding. This figure gives $\Delta G°$ values in kilojoules per mole; Ackers gives values in kilocalories per mole.

The total $\Delta G°$ in going from 2 mol of the unliganded dimer ($2\alpha\beta$) to the fully liganded tetramer ($\alpha_2\ \beta_2\ L_4$) should be the same, whether we proceed by way of $2\alpha\beta L_2$ as an intermediate or by way of the unliganded tetramer $\alpha_2\beta_2$. Experimentally this is true within fairly narrow limits of experimental error. The total $\Delta G°$ is -173.0 kJ mol^{-1} by one pathway and -172.7 kJ mol^{-1} by the other. The corresponding figures in terms of equilibrium constants are given at the bottom of the figure macromolecule bind the ligand to different extents. An excellent illustration is the work of Mills and Ackers (1979) and Ackers (1980) on the dimer–tetramer equilibrium in human haemoglobin and its linkage to the oxygen affinities of dimer and tetramer. We consider only the overall equilibria between the unliganded dimer ($\alpha\beta$) and tetramer ($\alpha_2\beta_2$) and the fully liganded forms $\alpha\beta L_2$ and $\alpha_2\beta_2 L_4$. The square diagram of Figure 5.18 shows the results. The dimer–tetramer association constant (K_A) for the unliganded molecules is some 45 000 times as great as the value (K_D) for the liganded molecules, corresponding to a more negative $\Delta G°$ for association of -26.2 kJ mol^{-1} (-59.7 versus -33.5). Correspondingly, $\Delta G°$ for the binding of four moles of O_2 to two moles of dimer is $2 \times -69.6 = -139.2$ kJ mol^{-1}, versus -113.3 kJ mol^{-1} for binding to one mole of tetramer. Ackers and his colleagues determined all the constants involved, including those for intermediate steps which we have not discussed here. (They also are in accord with the linkage equations.) However, even if we did not have all the data, we could (for instance) infer, from the greater oxygen affinity of dimers relative to tetramers, that the dimer–tetramer equilibrium must lie much further on the side of the tetramer in deoxy- than in oxyhaemoglobin. Conversely, if we knew the dimer–tetramer equilibrium values, we could infer the difference in oxygen affinities of the two forms. This exemplifies once again a point emphasized in Chapter 3: from one set of measurements one can often infer, by strict thermodynamic reasoning, the value of some other coefficient or equilibrium constant which may be more difficult to measure directly. The principles of thermodynamics enable us to recognize necessary relations between quantities which, on a casual inspection, might appear to be quite unrelated.

These data furnish an important example of what Wyman has termed polysteric linkage, involving the homotropic and heterotropic effects of ligand-linked dissociation and association in a macromolecule. For a general treatment of such linkage, see Colosimo, Brunori, and Wyman (1976). Wyman (1975) has also given a fundamental extension of the thermodynamic relations that link ligand binding phenomena to one another and to temperature, pressure, and other variables, making use of group theory.

CHAPTER 6

Calorimetry, heat capacity, and phase transitions

6.1 Calorimeters and their applications

In this chapter some selected applications of calorimetry to biological problems will be discussed and compared with other methods for the study of thermodynamic parameters of equilibria and transitions. We shall restrict ourselves to examples which can be interpreted in molecular terms and neglect the interesting field of study of cells or subcellular organelles. Some of the developments in this area have been discussed briefly in Chapter 1. Some chapters in a volume edited by Beezer (1980) give details of calorimetric measurements on cellular systems. Heat measurements on active muscle and nerve fibres have played an important role in the introduction of calorimetry into biology. The high resolution with respect to temperature and time, as well as the technical details of the methods used to study such systems, are described by Hill (1965) and Howarth (1970).

One might expect calorimetry to be the principal source of thermodynamic information, and it is indeed both a very versatile and sensitive technique. However, its use for research is still restricted to a relatively small number of laboratories, although the wider availability of commercial calorimeters has made it less of a prerogative of a few enthusiasts than it was 20 years ago. There are probably as many spectrophotometers, the biochemist's principal analytical tool, in a single major university as there are calorimeters in all the biochemical laboratories in the world; but the number of calorimeters is sure to increase greatly in the next decade or two.

The sensitivity and the quantity of the reactants required are, of course, closely related and determine the potential usefulness of a calorimeter to the study of a particular problem. In this section we wish to indicate some approximate magnitudes of these parameters in relation to values of ΔH which can be measured to given accuracies. Thermistors, thermocouples, and some semiconductor junctions can be used to obtain electrical signals which are functions of temperature or of the difference in temperature between a reference and reaction system. For considerable detail and further references to the technical aspects of calorimetry the reader should consult Wadsö (1970) and McGlashan (1979, Chapters 3 and 4). Specialized descriptions of methods used for microcalorimetric studies of biochemical reactions are found in a review by Langerman and Biltonen (1979). The procedures for the study of phase and structure transitions by differential scanning calorimetry are discussed by Krishnan and Brandts (1978) and Privalov (1974).

For biochemical calorimetry thermocouples or thermopiles (a number of thermocouples in series) are almost exclusively used as temperature sensors in one of two modes of operation: adiabatic direct measurements of temperature and changes due to the process of heat flow, and measurements in which the time integral of the heat conduction into or from a sink is recorded (Wadsö, 1970). Some practical points will be discussed in relation to the first of these approaches.

A convenient method of adiabatic calorimetry involves the use of a differential system with two well insulated reaction vessels, each containing one end of a thermopile and an electrical heater. The system can be used by recording the difference in temperature between the vessel with a reaction in progress and the reference vessel. Alternatively the temperature differential can be used to control the heater in one or the other of the vessels to zero the thermopile voltage. In the latter case the energy change can be calculated directly from the resistance of the heater and the current in coulombs per second, as derived on p. 17; this can be integrated over the time of the process. The heater can also be used to measure the thermal capacity and the heat loss of the reaction vessel. Clearly no system is perfectly adiabatic and corrections have to be made for this. If the temperature change during the reaction is to be used to calculate the enthalpy change, the specific heat of the reaction mixture has to be known at constant volume. This can be obtained from the electrical energy required to heat a known volume by 1 K. For most purposes this is not likely to be significantly different from 4.184 kJ l^{-1} K^{-1}, which is the energy required to raise the temperature of a litre of water. However, the smaller the volume of the reaction mixture the larger becomes the proportion of the heat taken up by the vessel.

Two types of chemical reactions can be used to calibrate calorimeters. The virtually instantaneous heat pulse obtained from mixing dilute solutions of NaOH and HCl results in $\Delta H = -55.85 \pm 0.1$ kJ mol^{-1}. For time-resolved enthalpy changes the OH^- ion-catalysed hydrolysis of ethyl acetate offers a suitable system. This reaction has an enthalpy change $\Delta H = -54.4$ kJ mol^{-1} and its pseudo-first-order rate constant can be varied easily over a wide range ($t_{1/2}$ from 1 to 100 s) by suitable changes in the concentration of excess KOH. Detail of heat measurements on this reaction are given by Johnson and Biltonen (1976) in their description of a heat conduction flow calorimeter.

A small thermopile with 10 copper–constantin junctions suitable for monitoring heat changes in a reaction vessel (with a volume of 0.1 dl) will give a change of approximately 0.4 mV K^{-1}. With modern microvoltmeters an accuracy of 0.1 μV is now not difficult to obtain. If the reaction under investigation has an enthalpy change of 50 kJ mol^{-1} the accuracy of 0.25×10^{-3} K results in a resolution with respect to concentration of about 2×10^{-5} M.

This means that the total reaction of a 2 mM solution can be measured with an accuracy of 1%. In theory only the molar concentration change and not the volume determines the temperature change. In practice the larger the reaction volume the smaller the corrections for the fraction of the heat taken up by the wall of the vessel and the thermopile itself.

Calorimetric observations of reactions are not only used for enthalpy measurements but also for the calculation of equilibrium constants, using heat production or uptake as a measure of the progress of a reaction as a function of time or concentration of reactants. In such cases the calorimeter is essentially used as an analytical instrument and the methods have found non-thermodynamic applications for clinical and industrial assays for reasons discussed below. For instance, when a ligand is titrated into a solution of a macromolecule the heat change will, after correction for the heat of mixing, be proportional to the extent of interaction. The heat taken up or evolved can be plotted against ligand concentration and analysed in a similar way to that applied to spectrophotometric data (see Sections 5.2, 5.8, and 5.9). Clearly every investigation benefits from being carried out by a range of methods. Some special property of a system may only be exhibited in the results obtained from one of the possible methods of observation.

There is an understandable preference for and advantage in methods which can be used to follow the progress of a reaction continuously as a function of time or reactant concentration, rather than by taking and analysing individual samples. Calorimetry can provide this for most systems, even to the extent that a special calorimeter has been designed for clinical and toxicological analysis (Goldberg, Prosen, Staples, Boyd, and Armstrong, 1975). This has proved very useful, especially for reactions which cannot be readily followed by optical changes. The use of different buffers as media for the reactions studied in a calorimeter can add to the information about the observed process. Two buffer systems most commonly used to study biochemical reactions are phosphate and tris(hydroxymethyl)/aminomethane (Tris). The former has a very small enthalpy of ionization, for the second pK proton uptake results in $\Delta H = -3.43$ kJ mol^{-1} while the proton uptake by Tris results in $\Delta H = -47.45$ kJ mol^{-1}. These facts can be used to design experiments involving reactions which are accompanied by proton uptake or release. First, such reactions can be followed by recording the large heat of ionization of the buffer and, secondly, the degree of change in ionization can be determined by following a reaction calorimetrically in two buffers, one with a small and another with a large heat of ionization. Calorimetric procedures have some advantages over the use of methods involving pH electrodes or indicators in that they can be carried out in buffered reaction mixtures and avoid some artefacts.

One of the most successful calorimeters for the study of biochemical reactions was designed by Buzzell and Sturtevant (1951). It was the first one to use the electronic recording techniques developed during World War II. Sturtevant's (1955) analysis of the heat of hydrolysis of *p*-nitrophenyl phosphate, from a study of the enzyme-catalysed reaction over a range of

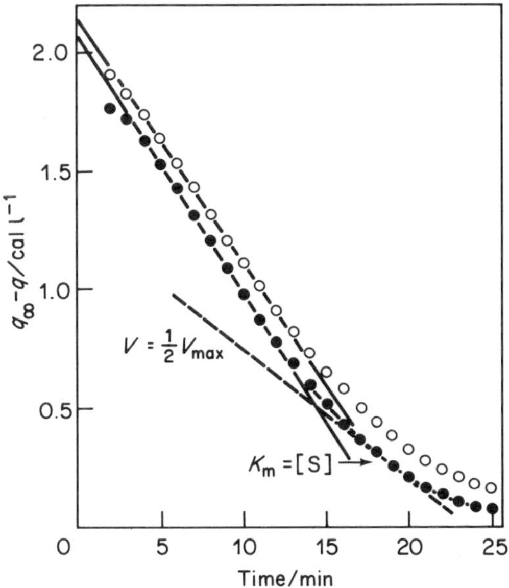

Figure 6.1 The kinetics of heat evolution in the hydrolysis of *p*-nitrophenyl phosphate at 25 °C by acid phosphatase. $[S]_0 = 0.43$ mM, pH = 5.27; •, calorimeter A; ○, calorimeter B

conditions, serves as a good example of the kind of information which can be obtained. Figure 6.1 shows the record of a typical experiment, which gives both the total heat evolution and its time course as catalysed by acid phosphatase. At low pH the reaction is essentially

$$O_2NC_6H_4OPO_3H^- + H_2O \rightarrow O_2NC_6H_4OH + H_2PO_4^-,$$

and for this transformation it was found that $\Delta H = -26.28$ kJ mol^{-1} at 25 °C. However, the reaction was carried out at different values of pH (3.62–5.85) in acetate buffer and, after corrections were made for the proton transfer in the following processes,

pK	Reaction			ΔH/kJ mol^{-1}
7.1	$O_2NC_6H_4OH + AcO^-$	$\rightarrow$	$O_2NC_6H_4O^- + AcOH$	23.0
6.9	$H_2PO_4^- + AcO^-$	$\rightarrow$	$HPO_4^{2-} + AcOH$	3.43
2.1	$H_3PO_4 + AcO^-$	$\rightarrow$	$H_2PO_4^- + AcOH$	−7.41

the enthalpy of dissociation and pK for the phosphate group of the substrate could be derived (see Figure 6.2). For the process

$$O_2NC_6H_4OPO_3H^- + AcO^- \rightleftharpoons O_2NC_6H_4OPO_3^{2-} + AcOH$$

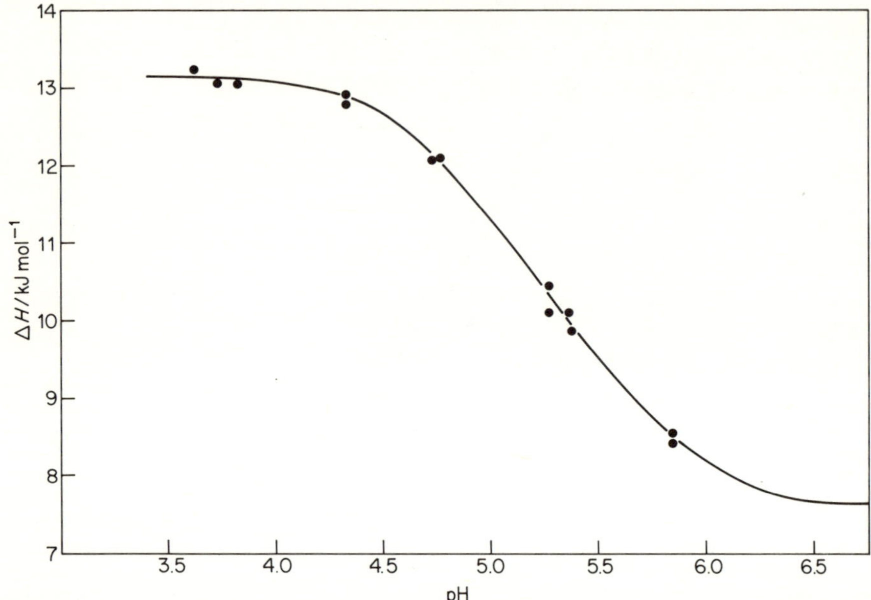

Figure 6.2 The pH dependence of the heat of hydrolysis of *p*-nitrophenyl phosphate at 25 °C. The change in ΔH over the pH range gives the heat of ionization of phosphate (see text). (Drawn from data from Sturtevant, 1955.)

we find that $\Delta H = -11.3$ kJ mol^{-1} and p$K = 5.4$. The enthalpy of dissociation of acetate can be considered negligible ($\Delta H = -0.38$ kJ mol^{-1}). Sturtevant (1955) compared the above with the enthalpy of hydrolysis of pyrophosphate at pH 7.3,

$$HP_2O_7^{3-} + H_2O \rightarrow H_2PO_4^- + HPO_4^{2-},$$

for which $\Delta H = -24.27$ kJ mol^{-1}, and a very similar value was obtained for

$$ATP^{3-} + H_2O \rightarrow ADP^{2-} + H_2PO_4^-.$$

An important event for the wider use of calorimetry was the appearance of commercial batch and flow calorimeters. Detailed technical specifications of these recent developments are given by Wadsö (1970). Modifications made by a number of investigators to suit their particular requirements are described by Langerman and Biltonen (1979) and Johnson and Biltonen (1976). An example of the use of the LKB batch calorimeter is illustrated in Figure 6.3. In these experiments 4.2 μl of ADP solutions of specified concentrations were injected into 4 ml of reaction mixture of both cells of a differential calorimeter. One cell contained the protein solution and the other cell only the solvent mixture. These experiments provided information about the enthalpy of binding (about -80 kJ mol^{-1}) and about the equilibrium of the binding process of ADP to myosin and to the catalytically active fragments of this protein (subfragment 1 and heavy meromyosin). Woledge (1977) described

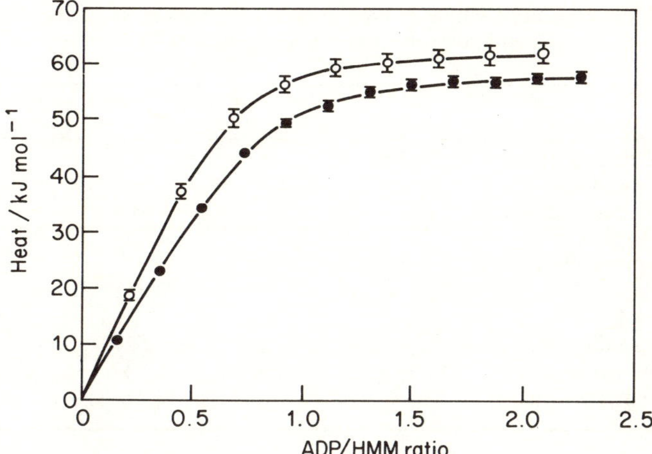

Figure 6.3 Heat produced during titration of HMM with ADP at 12 °C. The *points* show averaged results from five calorimetric titrations in 0.5 M KCl (•) and from six titrations in 0.1 M KCl (○). The quantities of HMM were 57.4 ± 9.1 (mean ± S.E.) and 46.2 ± 0.2 nmol in 4 ml of 0.1 M KCl and 0.5 M KCl buffer respectively. The amount of ADP added was 21.4 ± 0.2 nmol at each addition. The *bars* show ± S.E. The *lines* are binding curves fitted to the calorimetric results. (From Kodama, Watson, and Woledge, 1977.) HMM is heavy meromyosin.

these and similar measurements of enthalpy changes in interactions between myosin, actin, and nucleotides. The results of these studies are discussed in different contexts elsewhere in this volume (pp. 151 and 217). The thermodynamics of muscle contraction has been studied at every level of organization from the reactions of the pure isolated proteins to intact fibres contracting in response to electrical or mechanical stimulation. The obvious interest of energy balance between chemical and mechanical processes during contraction, relaxation, and isometric work encouraged many investigators to apply calorimetry to the study of temperature changes during various events taking place in muscle fibres (see Chapter 1 in this volume and Howarth, 1970). The interpretation of heat measurements on the chemical and mechanical events in muscle in terms of physiological processes is beyond our scope in this volume. A recent detailed summary of the results and outstanding problems has been presented by Homsher and Keen (1978).

6.2 Changes in heat capacity during reactions of proteins

In spite of the caution expressed earlier, thermodynamic quantities obtained from equilibrium or direct heat measurements are often used to propose and distinguish between molecular mechanisms. The Gibbs function is always obtained from equilibrium measurements and the relation $\Delta G^\circ = -RT \ln K$. The standard enthalpy change can be obtained either from equilibrium measurements at different temperatures, using the Gibbs–Helmholtz or the equivalent van't Hoff equation (see Section 4.11), or from direct calorimetry.

Table 6.1 Heat capacity changes in biochemical reactions at 25 °C: for further detail and references see Sturtevant (1977).

System	ΔC_p/J K^{-1} mol^{-1}
Heart lactate dehydrogenase + NAD^+ (binding)	− 351
Avidin + biotin (binding)	− 1 004
Aldolase + hexitol 1,6-diphosphate (binding)	− 1 715
α-Chymotrypsin unfolding (pH 7)	+12 887
Lysozyme unfolding (pH 7)	+ 6 527
Haemoglobin S gelation	− 837
Flagellin polymerization	−12 719

In either case care has to be taken to make sure that standard parameters refer to the standard states (see Chapter 3). Entropy changes of reactions, with the exception of phase transitions discussed in Section 6.3, are usually derived from differences between enthalpy and Gibbs energy data.

Of special interest in connection with molecular mechanisms are the changes in heat capacity accompanying chemical and physical processes. As pointed out in Section 4.11, the slope of a plot of ln K against $1/T$, giving $\Delta H^\circ/R$, will only be linear if the enthalpy change of the reaction studied is independent of temperature over the range investigated. The complex reactions involving large molecules in aqueous solutions usually result in changes in the heat capacity of the system. The relation

$$(\partial \Delta H^\circ/\partial T)_p = \Delta C_p \tag{6.1}$$

indicates that a change in heat capacity, at constant pressure, during a reaction will result in a non-linear plot of ln K versus $1/T$.

The relation between heat capacity and entropy is described formally in Section 2.3. A simplistic explanation of specific heat changes can be given by discussing an example. It has been stated (p. 36) that the transfer of non-polar groups of a protein from the hydrophobic interior into the aqueous phase can result in a decrease in entropy due to immobilization of water molecules at the increased hydrophobic surface area. As the temperature is increased the water structure on the hydrophobic surface 'melts'. Consequently, if the reaction (transfer of internal non-polar groups into an aqueous environment) is carried out at a higher temperature, there will be a smaller decrease in entropy. The change in entropy is related to a change in heat capacity by

$$T\left(\frac{\partial \Delta S}{\partial T}\right)_p = \Delta C_p. \tag{6.2}$$

The heat capacity is related to the number of degrees of freedom in the distribution of enthalpy states. As will be discussed below, changes in the heat capacity of a reaction often have compensating effects on the entropies and enthalpies of reactions, leaving the Gibbs energy nearly constant.

Changes in heat capacity during molecular interactions can be determined calorimetrically from equation (6.1) above, or calculated from the extended Gibbs–Helmholtz equation

$$T_1\Delta G^\circ(T_2) - T_2\Delta G^\circ(T_1) = \Delta H(0)(T_1 - T_2) - T_1T_2\Delta C_p \ln(T_1/T_2), \tag{6.3}$$

where $\Delta H(0)$ is the hypothetical enthalpy change at absolute zero (Sturtevant, 1974).

Some typical values for ΔC_p of reactions of proteins (see Table 6.1) show that unfolding of proteins results in large positive changes, which are ascribed to the exposure of hydrophobic groups to the aqueous environment. In contrast, most ligand binding processes result in negative values for ΔC_p, which are accounted for by a decrease in exposed surfaces, possibly due to tightening up of the structure. The binding of nucleotides to myosin, for instance, results not only in a large negative enthalpy change but also in a large negative change in heat capacity ($\Delta C_p = -1.4$ kJ K^{-1} mol^{-1}). This could help to characterize this important process (see also data tabulated in Section 4.9). Sturtevant (1977) defines six different causes of changes in heat capacity during reactions of proteins, which are obviously very complex.

Equations (6.1) and (6.2) show that a positive ΔC_p results in an increase in both ΔH and ΔS with increasing temperature. It is evident, therefore, that entropy and enthalpy changes with temperature can compensate to give fairly constant values for the Gibbs energy changes. For example the results obtained by Kodama and Woledge (1976) show that for ADP binding to myosin at 273.15 K $\Delta H = -57.1$ kJ mol^{-1} and $\Delta G^\circ = -33.58$ kJ mol^{-1} and that at 285.15 K $\Delta H = -73.1$ kJ mol^{-1} and $\Delta G^\circ = -33.21$ kJ mol^{-1}. The compensation is due to the difference in the standard entropy change, which is -86 J K^{-1} mol^{-1} at 273.15 K and -140 J K^{-1} mol^{-1} at 285.15 K.

The conclusions to be drawn from changes in heat capacity during such complex processes can be somewhat ambiguous. More information can be obtained if scanning calorimetry is employed (as indicated in the next section) to see whether any transitions occur over the relevant temperature, when each of the reactants and reaction products are examined on their own.

6.3 Differential scanning calorimetry and phase transitions

The distinct features of a scanning calorimeter are that the material to be investigated is heated at a constant rate and that the temperature is concomitantly recorded. The accuracy of the method is improved if such experiments are performed in the differential mode by comparing the enthalpy uptake of the material under investigation with that of a reference material such as the solvent subjected to the same energy input. It is essential that the energy input is sufficiently slow for equilibrium to be maintained at all times during an experiment (see Figures 6.4 and 6.5).

Differential scanning calorimetry has been used extensively to get information about a number of different events during structural and phase transitions of biological systems. Both of these phenomena are related to the melting of crystalline solids. The process of melting provides an excellent example for the illustration of the thermodynamics of cooperative transition. At the melting temperature T_m, the two phases are in equilibrium and the Gibbs energy for the transition

$$\text{crystal} \rightleftharpoons \text{liquid}$$

is zero. At T_m,

$$K = \frac{[\text{liquid}]}{[\text{crystal}]} = 1 \quad \text{and} \quad \Delta G^\circ = -RT_m \ln K = 0.$$

Hence

$$\Delta H = T_m \Delta S$$

and the heat of fusion ΔH_f is related to the entropy of fusion by

$$\Delta H_f / T_m = \Delta S_f.$$

During melting a quantity of heat equal to the latent heat of melting is taken up by the system at constant temperature. Similar events occur during transitions from the liquid to the gas phase, although that process does not illustrate the same kind of cooperativity. The removal of molecules from a crystalline lattice, as the kinetic energy ($\frac{1}{2}kT$) increases on heating, will result in a cooperative collapse of the crystal structure at the melting point. This is rather like removing a card from a house of cards.

Both ΔS_f and ΔH_f are positive – heat is taken up during the fusion process. Large molecules in solution, such as the helical structures of proteins and nucleic acids, also undergo cooperative transitions on heating. Individual molecules of regular structures behave like microcrystals. When one hydrogen bond of a turn in a helix breaks, then the others break more readily, and when one turn of a helix is broken the whole structure becomes more labile. Similarly the formation of a helical structure is a cooperative process; one turn in a helix can act as a seed for the formation of a helical molecule. Detailed and quantitative treatments of such processes are rather specialized and the reader is referred to Bloomfield, Crothers and Tinoco (1974, p. 343) and Flory (1969, p. 286) for further information on this subject.

The cooperative melting of a crystal or of a helical macromolecule results in the essentially simultaneous rupture of a large number of bonds. This in turn results in large positive values for ΔH and ΔS. Inspection of the equation

$$\Delta G = \Delta H - T\Delta S$$

shows that, assuming ΔH and ΔS are nearly independent of temperature, the $T\Delta S$ term determines the change in ΔG with temperature. The consequence of a large positive ΔS is that, in the region of T_m, a small change in temperature can change the Gibbs energy from a large positive to a large negative value.

The interest in phase transitions and in structure transitions, which can involve a small domain or the whole molecule, comes from their role in a range of biological systems. Such phenomena are also used for analytical purposes since they provide information about the structure, composition, and properties of proteins, nucleic acids, and the lipid bilayers of membranes. In most cases the transitions controlling biological phenomena are not induced by temperature changes, although some examples of such events will be given below. Alternative ways of triggering transitions are changes in the environment and molecular interaction which change the transition temperature to that of the organism. Examples of applications of scanning calorimetry to such problems are given in Sections 6.4 and 6.5.

Before a more detailed analysis is attempted of the form of phase transition illustrated in Figure 6.4, brief mention should be made of a classification sometimes used for these and related phenomena. In text-books of physics and physical chemistry (see for instance Atkins, 1978, p. 185) the 'order' of a phase transition is often defined in terms of the order of the temperature derivative of the chemical potential. If, as in the above example, $d\mu/dT$ is discontinuous, the transition is of first order; if $d^2\mu/dT^2$ is discontinuous, the transition is of second order; and so on. Rigorous theoretical treatments of the behaviour of higher order phase transitions are quite complex. An interesting and clear physical interpretation of such phenomena is given by Denbigh (1981, p. 207). Williams (1975) also discusses higher order transitions in his survey of phase changes in biological systems. At the time of writing, he was only able to give examples from inorganic chemistry. There is, however, no reason to believe that higher than first-order transitions will not be observed in lipids or macromolecules of biological interest (Radda, 1975).

Other topics covered in Williams' (1975) review on phase transitions will also be of interest to the reader since he shows the value of such studies for an understanding of the relation between structure, function, and biological cooperativity.

6.4 Calorimetric studies of stoicheiometry

The classical study of Brown and Hill (1923) using thermodynamic data in an attempt to determine the stoicheiometry of oxygen binding to haemoglobin is of historical interest. It was known from iron analysis that haemoglobin contained one equivalent of haem per 16.7 kg of protein. Haemoglobin was thought to be an aggregate of units of that size. Only the stoicheiometry of oxygen binding was known, but not the oligomeric molecular weight. Brown and Hill obtained values for the heat of oxygen binding calorimetrically per gram of oxygen and the molar enthalpy from application of the van't Hoff equation, using the temperature dependence of the equilibrium constant.

Their aim was to investigate the order of the reaction

$$Hb + nO_2 \rightarrow Hb(O_2)_n$$

with respect to the oxygen concentration, since it had been found that S-shaped binding curves could be represented by the equation

$$y/(1 - y) = K[O_2]^n$$

(see p. 197). They determined the heat of the reaction of oxygen binding in terms of

$$q/\text{g of oxygen bound} = q_g$$

and the heat calculated from the temperature dependence of the equilibrium binding constant, ΔH_{vH}. The ratio

$$\Delta H_{vH}/q = n \times 32.$$

They found values for n in the region of 2.3. It was subsequently established that haemoglobin is a tetramer under physiological conditions and binds four oxygen molecules with cooperative interactions between the binding sites (see Section 5.8).

Comparisons of van't Hoff and calorimetric enthalpies have since been applied to the interpretation of many cooperative phenomena. Both enthalpies can be derived from appropriate measurements in a differential scanning calorimeter.

The enthalpy uptake of a system heated from a solid at $T = 0$ to a liquid at T' is described by

$$H_L(T') - H_S(0) = \int_{T=0}^{T_m} C_p^S(T)\, dT + [H_L(T_m) - H_S(T_m)] + \int_{T_m}^{T'} C_p^L(T)\, dT,$$

where the term in square brackets is the heat of fusion.

Figure 6.4 shows diagrams of enthalpy and specific heat as a function of temperature across a phase transition. In this idealized illustration the specific heat is the same before and after the transition. In most real cases some changes in the value and the temperature derivative of the specific heat occur (see Figure 6.5).

If we observe a two-state transition A $\leftrightarrow$ B with no overall change in heat capacity (as in Figure 6.4), we obtain information about the heat uptake as a function of α, the extent of the reaction. T_m is the temperature when $\alpha = 0.5$. Sturtevant (personal communication) has suggested the following analysis of data obtained from calorimetric measurements on such systems. A tangent drawn at T_m defines T_1 and T_2 as follows:

$$d\alpha/dT = 1/(T_2 - T_1).$$

The equilibrium constant for a process with no change in mole numbers is

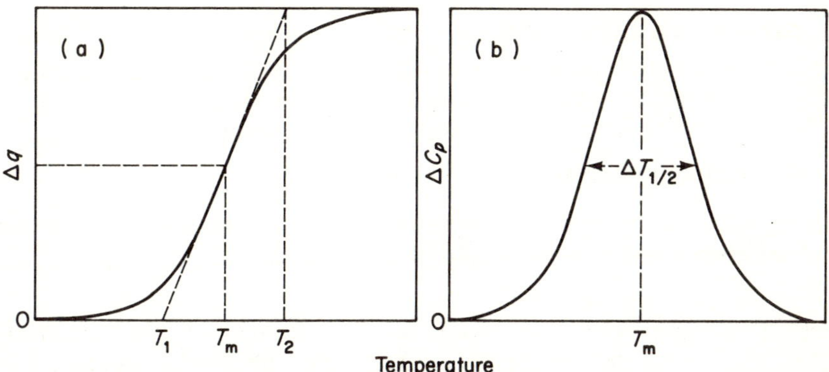

Figure 6.4 The variation with temperature of the excess enthalpy (a) and the excess specific heat (b) during a two-state process in which there is no overall change of heat capacity. (From Sturtevant, 1974.)

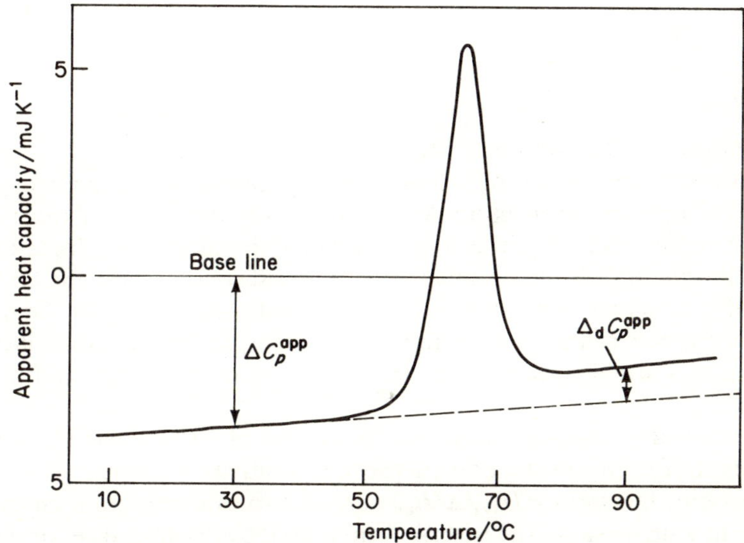

Figure 6.5 An example of scanning microcalorimetric recording of apparent heat capacity of a dilute protein solution on heating over a broad temperature range. (Converted to joules from Privalov, 1979.)

$$K = \alpha/(1 - \alpha),$$

so that

$$\frac{d(\ln K)}{dT} = \left[\frac{1}{\alpha(1 - \alpha)}\right] \frac{d\alpha}{dT} = \frac{\Delta H_{vH}}{RT^2} \cdot \tag{6.4}$$

It follows from $\alpha = 0.5$ at T_m that the van't Hoff enthalpy can be expressed as

$$\Delta H_{vH} = 4RT_m^2/(T_2 - T_1).$$

Figure 6.4(b), the derivative of the enthalpy uptake with respect to temperature, represents the excess heat capacity, sometimes called the heat capacity anomaly, during the transition. The maximum $(C_{ex})_{max}$ occurs at $\alpha = 0.5$ and it follows that

$$C_{ex}/(C_{ex})_{max} = 4\alpha(1 - \alpha)\left(\frac{T_m}{T}\right)^2.$$

Since T is approximately equal to T_m throughout the experiment, $C_{ex}/(C_{ex})_{max} = 0.5$ at $\alpha \approx 0.147$ and 0.853. On substitution of these values into the integrated form of equation (6.4),

$$\ln\left(\frac{1 - \alpha}{\alpha}\right) = \frac{\Delta H_{vH}}{R}\left(\frac{1}{T} - \frac{1}{T_m}\right)$$

one obtains

$$\Delta H_{vH} \approx 6.9\ T_m^2/\Delta T_{1/2}$$

for reactions with T_m around 350 K.

For two-state processes $\Delta H_{vH}/\Delta H_{cal}$ must equal gram equivalents of the perfectly cooperative transition A to B (ΔH_{vH} is the heat per mole of the cooperative unit and ΔH_{cal} the heat per gram of the substance). It is clear that the sharper the transition the smaller is $\Delta T_{1/2}$ and consequently the larger the group of interacting units obtained from the ratio $\Delta H_{vH}/\Delta H_{cal}$. Alternative methods for the calculation of thermodynamic parameters from measurements in a scanning calorimeter are described by Krishnan and Brandts (1978).

In Sturtevant's analysis and in the discussion of Brown and Hill's (1923) experiments for the evaluation of the cooperativity of oxygen binding to haemoglobin, the ratio $\Delta H_{vH}/\Delta H_{cal}$ was used to evaluate gram equivalents reacting in a single step. The inverse ratio is used by Privalov (see for instance Privalov, 1979) and others to evaluate the fraction of a molecule of known molecular weight which reacts in each step.

The above arguments, applied to phase transitions during melting of solids, give information about the degree of interaction between molecules. This has important applications to events in the lipid environments of cells to be discussed below. Transitions within macromolecules can be interpreted in an analogous way by considering different parts of the macromolecule as units which may or may not interact in a process such as helix to random coil transitions. The thermodynamic analysis of such processes is a complex and specialized problem, which usually only has approximate solutions (Sturtevant, 1974). However, many useful indications about the behaviour of systems of biological interest are obtained by the use of differential scanning calorimetry. Some of these studies are worthy of discussion here, although they do not give strictly thermodynamic information.

Thermal transitions of pure proteins in solution can show whether such molecules denature in a single step (two-state process) or whether a number of domains 'melt' at different temperatures. An example of a two-state process is the thermal unfolding of lysozyme in acidic solutions (see Figure 6.5). A single transition is observed under each condition and comparison of ΔH_{vH} and ΔH_{cal} shows that the whole molecule undergoes a single cooperative transition (two-state system). Privalov (1979) plotted the ratio $\Delta H_{cal}/\Delta H_{vH}$, taking ΔH_{cal} for 14.4 kg, against the transition temperature of lysozyme in solutions of different pH. The ratio did not exceed 1.1, indicating a two-state transition under all conditions investigated. The record in Figure 6.5 also shows that the heat capacity of the unfolded protein is greater than that of the native protein. This can be due both to the greater flexibility and the exposure of hydrophobic areas.

In contrast to the results obtained with lysozyme, Tiktopulo and Privalov (1978) showed that the thermal transition of papain does not represent a two-state system. They find that the analysis of the records of differential calorimetric investigations give ratios of about 0.56 for $\Delta H_{cal}/\Delta H_{vH}$, indicating that two domains of the molecule undergo independent transitions. Similar results described by Mateo and Privalov (1981) indicate that pepsinogen, which undergoes a nicely reversible temperature transition, also unfolds in two independent domains. In both these cases the multistate transition occurs over a single temperature range. However, when pepsinogen is converted to pepsin, the two domains unfold at different temperatures, as shown in Figure 6.6. Privalov (1974, 1979) and Krishnan and Brandts (1978)

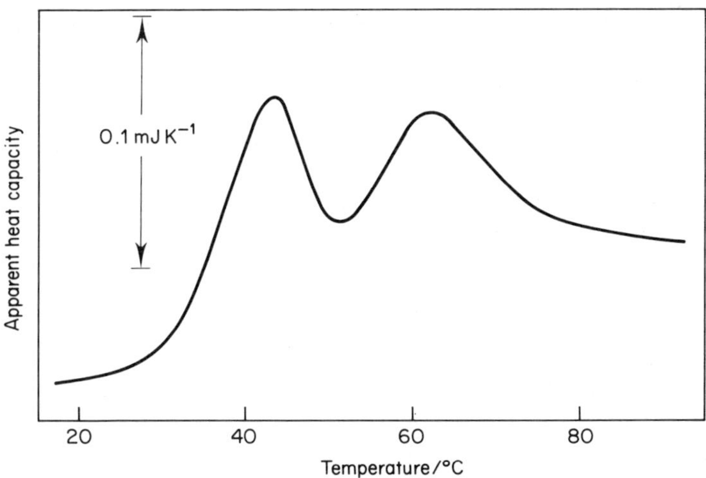

Figure 6.6 Calorimetric recording of heat absorption during denaturation of pepsin. The solution contained 1.42 mg ml^{-1} protein, 5 mM sodium phosphate, 0.1 M NaCl (pH 6.5). (Modified from Mateo and Privalov, 1981.)

give other examples of two-state transitions in proteins and nucleic acids. If a protein undergoes several transitions at different temperatures, this can be due either to different parts of the molecule unfolding at different temperatures or to a multiplicity of transitions of the whole molecule. Detailed analysis of the data permits distinction between the two possibilities (see for instance Privalov, 1982).

6.5 Application of optical techniques to the study of transitions

Privalov (1974), who designed one of the most sensitive scanning microcalorimeters, expressed the following opinion about its application: 'The possibility of direct measurement of the most important thermodynamic parameters is, naturally attractive. Primarily it is of interest to those dealing directly or indirectly with the problems of intramolecular cooperative transformations of biopolymers, the stability of their space structures and the intramolecular interactions with the solvent which ultimately determine the structures and structural transformations of biopolymers in solution. To this should be added the importance of the study of transitions of the media in non-aqueous environments, namely the lipids of the membrane phase which control many important biological events'. However, optical methods can also be used.

Studies of thermal transitions in DNA solutions have provided a considerable amount of useful information. It should be pointed out that in many cases the melting of nucleic acids can be followed by observation in a spectrophotometer programmed to record extinction at 258 nm as a function of increasing temperature. The melting (or unwinding) of the double helix results in the disappearance of the characteristic hypochromic shift of the base-paired system (see Figure 6.7). An empirical relation has been found between the melting point (T_m in degrees Celsius) and the ratio of guanine–cytosine and adenine–thymine base pairs. The former (GC) have three hydrogen bonds and

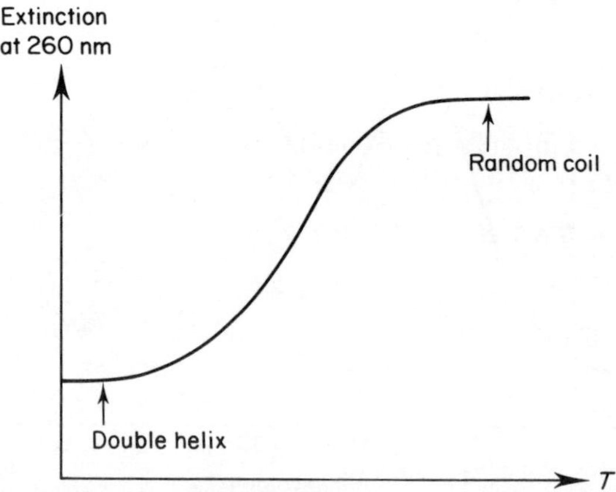

Figure 6.7 The increase in extinction at 260 nm shows the disappearance of hypochromicity on melting of DNA. The effect of composition of DNA on the melting point is discussed in Section 6.5

the latter (AT) two hydrogen bonds in the paired double helix. The expressions used are

$$T_m = 69 + 41f_{GC} \qquad \text{for DNA},$$

$$T_m = 62 + 78f_{GC} \qquad \text{for RNA},$$

where f_{GC} is the fraction of GC pairs. (Note that in RNA uracil replaces thymine.) Another approximate relation between melting point and composition, which has been found useful, concerns the estimate of homology between two strands of a duplex. If one wishes to know the number of base replacements between two genes of related species, one compares the melting points of the homoduplexes with that of the artificially formed heteroduplex of the two samples of DNA. The difference, ΔT_m in degrees Celsius, corresponds to the percentage substitution.

Many processes involving DNA are controlled by the unwinding and helix–random coil transition of the double helix. This in turn is much affected by the ionic composition of the medium. Studies of melting in different media give information about the energy requirement for unwinding DNA under different conditions. It is now known that enzymes utilize the energy available from ATP hydrolysis to mediate structure changes in DNA (Gellert, 1981).

In many other cases transitions in macromolecules can be studied by recording some optical property (transmission, fluorescence, or optical rotation) as a function of temperature. An increasing number of new techniques are applied to the study of such processes. In such investigations the equilibrium constant

$$K = f_u/f_n,$$

where f_u is the fraction unfolded and f_n is the fraction of native molecules, and its temperature dependence are used to evaluate the thermodynamic parameters. If we again define T_m as the temperature where $K = 1$, then at that temperature

$$\Delta G = 0 \quad \text{and} \quad T_m\Delta S = \Delta H,$$

and for ΔG at a temperature T (near T_m), assuming that ΔH and ΔS are independent of T,

$$\Delta G = \Delta H - T\Delta S = \Delta H\,(1 - T/T_m).$$

Using the approach of Elwell and Schellman (1977) one can derive the parameters for the transition of state n to state u as follows. The specific heat change is assumed to be constant over the temperature range of studies of biopolymers and ΔH follows the linear law

$$\Delta H = a + bT,$$

where $b = \Delta C_p$ for the transition. Under given conditions

$$\Delta H = \Delta H_m + b(T - T_m),$$

where ΔH_m is the enthalpy change at the midpoint of the transition. It follows that

$$\left(\frac{\partial \Delta S}{\partial T}\right)_p = \frac{1}{T}\left(\frac{\partial \Delta H}{\partial T}\right)_p = \frac{b}{T}$$

Elwell and Schellman derive from this that

$$\Delta S = \Delta S_m + b \ln(T/T_m)$$

and

$$\Delta G = \Delta H_m (1 - T/T_m) - b[(T_m - T) + T \ln (T/T_m)]$$

and

$$\Delta C_p = \frac{\partial}{\partial T}\left(R\frac{\partial(\ln K)}{\partial(1/T)}\right).$$

From such equilibrium studies ΔH_{vH} can be obtained. Although ΔH_{cal} would have to be obtained separately for any firm conclusion about the nature of the transition (two-state or multistate), it is argued that large values for ΔH_{vH} are evidence for two-state transitions.

6.6 Phase transitions in aqueous lipid systems

A large area of investigations of phase transitions using scanning calorimetry concerns lipids. A considerable amount of information has been accumulated about pure lipids and about various natural membranes or the total lipids extracted from them. The temperatures at which phospholipids have transitions from the solid to the fluid phase are dependent upon chain length, degree of saturation, and the presence of other components, particularly cholesterol. The effect of chain length is illustrated in Figure 6.8. The transitions discussed here are those of lipids in suspension in dilute aqueous systems. The method of suspension has some effects (Sturtevant, 1974). In sonicated suspensions of highly purified lipids very sharp transitions are observed and cooperative interactions of upwards of 100 molecules are observed in suspensions of multilamellar diacylphosphatidylcholine lipids with saturated hydrocarbon chains. Unsaturated hydrocarbon chains have many isomers and form less ordered solids which cause lowering of melting points. Many membrane functions can only occur in the fluid state and animals which function at low temperatures tend to have polyunsaturated lipids. The melting point of lauric acid (C_{12}) is 43.5 °C, while that of linderic acid (C_{12} with one double bond) is about 1 °C. The presence of cholesterol, a frequent component of animal lipid mixtures, broadens and lowers the melting point of the system.

The results of several investigations gave evidence for the essentially fluid nature of functional membranes. Cherry (1979) reviews a number of

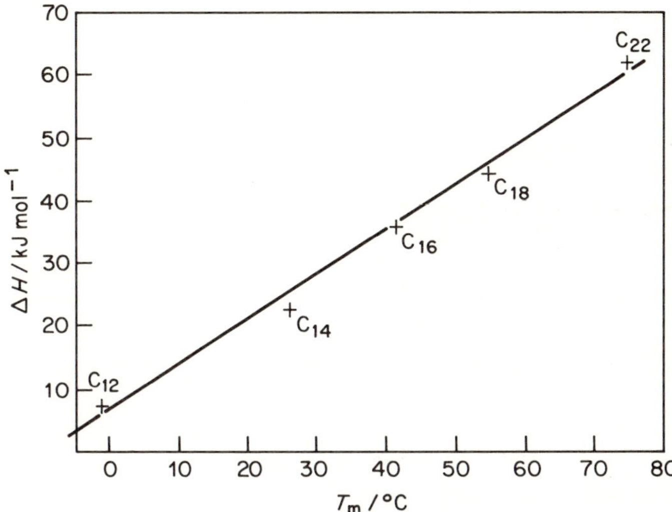

Figure 6.8 Enthalpies for the upper transition of phosphatidylcholines with different acyl groups plotted against transition temperatures. (Data from Ladbrooke and Chapman, 1969.)

techniques which demonstrate the rapid diffusion of proteins in membrane lipids. In microorganisms the composition of membrane lipids can be changed by mutations or (in auxotrophs) by changing the lipids in the growth medium (Linden and Fox, 1975). In both cases lowering of the temperature for phase transitions permitted organisms to survive at lower temperatures.

Evidence for fluidity of membranes from measurements of diffusion rates of lipid molecules or specific proteins embedded in the double layer was referred to above. The techniques used involve labelling of molecules with free radical or fluorescent markers. Discussion of these kinetic approaches would take us even further from thermodynamic principles, from which we have already strayed in this chapter. It is, however, of interest to state that intrinsic membrane proteins partition preferentially into the fluid phase of lipids and that lipids exchange rapidly horizontally in the monolayer of a double layer, but interchange between the two layers is *very* slow. Protein molecules have rotational diffusion constants in the double layer which correspond to lipid viscosities which are approximately 10^3 times that of water. For further information about phase transitions and fluidity characteristics in membranes, the reader should consult Chapman (1975). References to the application of other techniques to the study of such transitions will also be found there.

APPENDIX

Some guidelines to data handling

In many scientific investigations experiments are designed to test a hypothesis or to decide between two hypotheses. The measurements have to be carried out using instruments which are accurate enough, with respect to critical variables, to produce results of sufficient precision to allow for clear decisions. If we want to know whether one or two calcium ions bind to a particular protein molecule the final precision of data need not be better than ±10%. This, however, would be quite inadequate if we wish to decide between nine or 10 binding sites. If one carries out investigations into the Gibbs energy of the formation of ATP from ADP and orthophosphate, one is well advised to survey the accuracy required for decisions on current problems in the energy balance of metabolic and other processes coupled to its formation and utilization. It is rarely justified to go to the necessary expenditure of time and material resources to extend the presently required accuracy by more than a small factor. If more accurate data are required at some future date, earlier measurements are rarely found to have been taken under the right conditions.

The *Oxford English Dictionary* does not distinguish between accuracy and precision and the two words are used synonymously in everyday speech. However, in scientific measurements a distinction is made. Accuracy is the closeness of a measured or computed value to its true value. Precision is the closeness of repeated measurements of a quantity with the same instrument. A biased but sensitive scale will yield precise but inaccurate data. One can make a precise statement – the heat of fusion of ice is 2.346 J mol^{-1} – but it is inaccurate! (in the sense that it is incorrect; the real value is 6.00 kJ mol^{-1}).

When assessing the accuracy of an experimental measurement one always has to look for the weakest link; that is, if a result is dependent on weighing a substrate and on measuring a temperature, one has to ask which of the two will contribute most to the inaccuracy of the final result. This may sound obvious or trivial, but it is an essential consideration when one wishes to improve the accuracy of the overall result of an investigation.

In the present brief discussion of data analysis we have to omit on the one hand the elementary statistical calculations of means, standard deviations, and standard error of the mean and on the other hand more advanced and detailed statistical analysis leading to quantitative estimates of confidence limits and significance tests. It is hoped that the inclusion, in the present volume, of methods which can help with the numerical analysis of one's own data and those from the literature will be found a convenience. Thermodynamics is a science which deals largely with the interpretation of experimental data.

There is a very large number of very good books on statistical analysis and on numerical methods at every level of sophistication; it is not our intention to make an amateur's attempt to compete with these. Several texts will be referred to for particular methods. However, since we believe that the proper planning and numerical analysis of experiments is an essential part of the interpretation of thermodynamic data, we wish to present a brief account of some methods which we have found useful. These procedures will also demonstrate how results should be presented so that they can be used by other investigators for interpretation in terms of different models. The methods will be discussed with examples from the field of ligand binding, but they should be generally applicable to fitting data to any of the equations discussed in this volume. The procedures can, of course, be used to analyse published results which one wishes to examine critically, as well as those coming from one's own experiments.

We started with the interpretation in terms of linear equations and we shall use one of these to make a number of points. Figure A.1 shows a Scatchard plot of $\bar{\nu}/[\mathrm{L}]$ against $\bar{\nu}$. The equation

$$\bar{\nu}/[\mathrm{L}] = Kn - K\bar{\nu}$$

can be treated by the linear regression method in terms of $y = a + bx$, where $y = \bar{\nu}/[\mathrm{L}]$, $x = \bar{\nu}$, $b = K$, and $a = Kn$. By the method of 'least squares deviations' for the calculation of the best linear fit one can derive an expression for the parameters a and b as follows:

The condition for least squares requires $(y_{\mathrm{obs}} - y_{\mathrm{calc}})^2$ to be minimized. This can be written as

$$\sum (y - a - bx)^2 \text{ is minimum}$$

if

$$S = \sum y^2 - 2b \sum xy - 2a \sum y + 2ba \sum x + b^2 \sum x^2 + Na^2,$$

where N is the number of observations (or points on our graph). The minimum for S can be found by differentiation with respect to a at constant b and with respect to b at constant a, and setting both partial differentials equal to zero:

$$\left(\frac{\partial S}{\partial b}\right)_a = -2\sum xy + 2a\sum x + 2b\sum x^2 = 0,$$

$$\left(\frac{\partial S}{\partial a}\right)_b = -2\sum y + 2b\sum x + 2Na = 0;$$

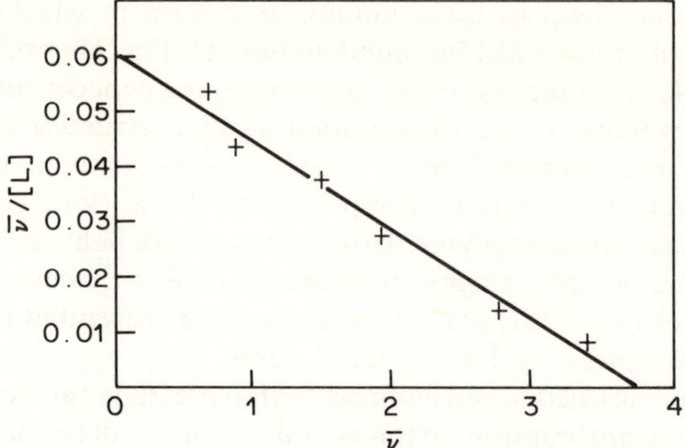

Figure A.1 A numerical example of applying the least squares method to the equation $\bar{\nu}/[L] = Kn - K\bar{\nu}$:

$\bar{\nu}$	$[L] \times 10^6$	$(\bar{\nu}/[L]) \times 10^{-6}$
0.670	12.5	0.0536
0.874	20.0	0.0437
1.503	40.0	0.0376
1.934	70.0	0.0276
2.787	200	0.0139
3.421	400	0.0086

Comparing this equation with the equation $y = ax + b$, where $y = \bar{\nu}/[L]$ and $x = \bar{\nu}$, we obtain by linear regression, using equations (A.1)–(A.3), $a = -K = -1.60 \times 10^4$ and $b = Kn = 6.06 \times 10^4$. Thus $K = 1.60 \times 10^4\ \text{M}^{-1}$, $n = -b/a = 3.80$ and $r^2 = 0.974$.

thus

$$a \sum x + b \sum x^2 = \sum xy,$$

$$aN + b \sum x = \sum y.$$

Using Cramer's rule we obtain

$$a = \frac{\sum xy \sum x - \sum y \sum x^2}{(\sum x)^2 - N \sum x^2}, \tag{A.1}$$

$$b = \frac{\sum x \sum y - N \sum xy}{(\sum x)^2 - N \sum x^2}. \tag{A.2}$$

A numerical example of applying the linear least squares method is shown in Figure A.1.

This form of statistical analysis is widely used in spite of the fact that it has a number of shortcomings. One important problem is that the simple analysis of the equation $y = a + bx$ assumes that the dependent variable is only contained in y, while x should only contain the independent variable which is supposedly error free. This would apply in binding equations in which the observed concentration of the complex (dependent variable) occurs exclusively on the left-hand side of the equation, while only the ligand concentration occurs as a variable (independent) on the right-hand side; a and b are called the parameters to be determined by the variables. Unfortunately linear forms of binding equations with the dependent variable on one side only have other disadvantages for statistical analysis due to the fact that this variable then appears as a reciprocal. These problems have been discussed in detail by Colquhoun (1971, p. 257) and, in relation to substrate binding in enzyme reactions, by Roberts (1977, p. 285).

The problem of cases when both variables are random variables, that is neither of them are considered to be free from error, can be treated by calculation of correlation coefficients. Details of this approach, with some examples, can be found in Bulmer (1979, p. 73 and 221). If the regression of y on x is expressed as $y = a + bx$ and the regression of x on y is expressed as $x = a' + b'y$, then the square of the correlation coefficient is

$$r^2 = bb' = \frac{(\sum x \sum y/N - \sum xy)^2}{(\sum x^2 - (\sum x)^2/N)(\sum y^2 - (\sum y)^2/N)} \,. \tag{A.3}$$

If x and y are perfectly correlated, $b = 1/b'$ and $r^2 = 1$. As an aside it is worth noting that the above expression for r^2 and equation (A.2) for b are in a form most suitable for programmable calculators which usually have only enough stores for $\sum x$, $\sum y$, $\sum xy$, and the square terms needed. If a calculator or computer is used which can store x and y for all the data points one can use the expressions

$$b = \frac{\sum x \sum y/N - \sum xy}{\sum (x - \bar{x})^2} \,, \tag{A.4}$$

$$r^2 = \frac{(\sum x \sum y/N - \sum xy)^2}{\sum (x - \bar{x})^2 \sum (y - \bar{y})^2} \,. \tag{A.5}$$

The equivalence of equations (A.3) and (A.5) is readily shown:

$$\sum (x - \bar{x})^2 = \sum (x^2 - 2x\bar{x} + \bar{x}^2) = \sum x^2 - 2\bar{x} \sum x + N\bar{x}^2;$$

since $\sum x = N\bar{x}$ it follows that

$$\sum x^2 - 2N\bar{x}^2 + N\bar{x}^2 = \sum x^2 - N\bar{x}^2$$

and

$$\sum (x - \bar{x})^2 = \sum x^2 - (\sum x)^2/N.$$

Since r^2, as defined above, must lie between 0 and 1, it follows that r lies between 1 and -1. The correlation coefficient is negative when b is negative. We therefore have a reasonable measure of linear correlation. It is perfect when $r = 1$ or -1 and there is no linear correlation as r approaches zero. For a quantitative treatment of the relation between r and confidence tests the reader must be referred to Bulmer (1979) or other specialized statistics texts.

An important practical point emphasized by Dowd and Riggs (1965) is the necessity to collect data over a wide range of ligand concentrations corresponding to a minimum range of saturation (liganding) from 0.1 to 0.9. If deviations occur from the simple form of binding to identical and independent sites and therefore from the hyperbolic equation, data have to be available at lower and higher values for degree of liganding to obtain a correct picture.

Many discussions of the statistical analysis of experimental data emphasize the necessity of weighting data points, which means that the points in regions of the curve which are likely to yield more accurate data are given a greater weight than those from other parts of the curve (Colquhoun, 1971, p. 272). For this purpose one needs to determine the variance (see Colquhoun, 1971) of each point, or at least of each of several sections of the curve. One cannot predict whether data will be more accurate at any one particular level of saturation rather than another. The inspection of continuous records now commonly used is not as suitable for this purpose as the repetitive determination of saturation at each value for [L].

The linear least squares method can also be used to analyse the relations between the equilibrium constant and temperature (van't Hoff equation, see p. 152 and Section 4.3),

$$\ln K = -\frac{\Delta H^\circ}{RT} + C, \tag{A.6}$$

which is widely used in this volume for the evaluation of the heats of different processes. In this case the dependent variable K is only on the left-hand side and only the independent variable occurs in a reciprocal form. In principle this linear relation can be determined from two accurate points. One may, therefore, determine the equilibrium constant several times at each of two temperatures, calculate the mean at both temperatures, and calculate the slope of the line through the two averaged points. However, one also wishes to know whether one is really dealing with a linear relation or whether ΔH° changes with temperature (see p. 126). For this purpose one needs fairly evenly distributed data over the whole temperature range so that a meaningful correlation coefficient can be calculated.

A linearized equation, whether for an equilibrium constant (hyperbola) or the van't Hoff relation (exponential), does not provide as critical a test as a fit to the original non-linear equation. The advent of hand-held programmable calculators and microcomputers, and their availability in most laboratories, makes it possible to fit raw data directly to the non-linear equation without access to or understanding of statistical programmes of mainframe computers. A worked example of fitting data to the equation

$$[\mathrm{BL}] = \frac{[\mathrm{B}]_o[\mathrm{L}]}{[\mathrm{L}] + K^{-1}} \tag{A.7}$$

should illustrate the method which can be applied equally well to exponential equations. This form of the binding isotherm is suitable for the analysis of data from equilibrium dialysis or electrode titration experiments. It relates the concentration of occupied sites [BL] to ligand concentration [L] and enables one to evaluate the two parameters, the concentration of binding sites $[B]_o$, and the association constant K.

The differential correction technique (see for example McCalla, 1967, p. 256) depends on the solution of a matrix of the following type:

$$\begin{bmatrix} \Sigma\left(\frac{\partial f}{\partial C_1}\right)^2 & \Sigma\frac{\partial f}{\partial C_1}\frac{\partial f}{\partial C_2} & \cdots & \Sigma\frac{\partial f}{\partial C_1}\frac{\partial f}{\partial C_m} \\ \Sigma\frac{\partial f}{\partial C_2}\frac{\partial f}{\partial C_1} & \Sigma\left(\frac{\partial f}{\partial C_2}\right)^2 & \cdots & \Sigma\frac{\partial f}{\partial C_2}\frac{\partial f}{\partial C_m} \\ \vdots & \vdots & & \vdots \\ \Sigma\frac{\partial f}{\partial C_m}\frac{\partial f}{\partial C_1} & \Sigma\frac{\partial f}{\partial C_m}\frac{\partial f}{\partial C_2} & \cdots & \Sigma\left(\frac{\partial f}{\partial C_m}\right)^2 \end{bmatrix} \begin{bmatrix} \Delta C_1 \\ \Delta C_2 \\ \vdots \\ \Delta C_m \end{bmatrix} = \begin{bmatrix} -\Sigma\frac{\partial f}{\partial C_1}R_i \\ -\Sigma\frac{\partial f}{\partial C_2}R_i \\ \vdots \\ -\Sigma\frac{\partial f}{\partial C_m}R_i \end{bmatrix} \tag{A.8}$$

This relates to the expression for the dependent variable y being a function of the independent variable x and the parameters C_1 to C_m:

$$y = f(x,\ C_1, C_2, \ldots, C_m).$$

To start the iterative procedure one uses a set of equations with estimated parameters and the solution gives the corrections ΔC_1 to ΔC_m for the parameters. The corrected parameters are used for the next set of equations and so on until the ΔC values become negligible. R_i is obtained from the x and y of each point from

$$R_i = f(x,\ C_1^0, C_2^0, \ldots, C_m^0) - y \tag{A.9}$$

for the first estimate and from

$$R_i = f(x,\ C_1^1,\ C_x^1,\ \ldots,\ C_m^1) - y$$

$$C_1^1 = C_1^0 + \Delta C_1, \ \ldots\ C_m^1 = C_m^0 + \Delta C_m$$

for the first correction and so on for further corrections of the parameters.

For the case of fitting data to equation (A.7) one has to evaluate, for successive cycles of corrections, the partial differentials

$$\left(\frac{\partial[\mathrm{BL}]}{\partial[\mathrm{B}]_o}\right)_K = \frac{[\mathrm{L}]}{K^{-1} + [\mathrm{L}]}$$

and

$$\left(\frac{\partial[\mathrm{BL}]}{\partial K^{-1}}\right)_{[\mathrm{B}]_\mathrm{o}} = \frac{-[\mathrm{B}]_\mathrm{o}[\mathrm{L}]}{(K^{-1} + [\mathrm{L}])^2}$$

as well as

$$R_i = \{[\mathrm{B}]_\mathrm{o}[\mathrm{L}]/(K^{-1} + [\mathrm{L}])\} - [\mathrm{BL}]$$

for each data point relating [BL] to [L], and sum them for evaluation of $\Delta[\mathrm{B}]_\mathrm{o}$ and ΔK to give the matrix form of the equation. For the simple case of a function with two parameters the solution is obtained by Cramer's rule:

$$\begin{bmatrix} \sum \left(\frac{\partial[\mathrm{BL}]}{\partial[\mathrm{B}]_\mathrm{o}}\right)^2 & \sum \frac{\partial[\mathrm{BL}]}{\partial[\mathrm{B}]_\mathrm{o}} \frac{\partial[\mathrm{BL}]}{\partial K^{-1}} \\ \sum \frac{\partial[\mathrm{BL}]}{\partial K^{-1}} \frac{\partial[\mathrm{BL}]}{\partial[\mathrm{B}]_\mathrm{o}} & \sum\left(\frac{\partial[\mathrm{BL}]}{\partial K^{-1}}\right)^2 \end{bmatrix} \begin{bmatrix} \Delta[\mathrm{B}]_\mathrm{o} \\ \Delta K^{-1} \end{bmatrix} = \begin{bmatrix} -\sum \frac{\partial[\mathrm{BL}]}{\partial[\mathrm{B}]_\mathrm{o}} R_i \\ -\sum \frac{\partial[\mathrm{BL}]}{\partial K^{-1}} R_i \end{bmatrix},$$

$$\Delta[\mathrm{B}]_\mathrm{o} = \frac{-\sum \frac{\partial[\mathrm{BL}]}{\partial[\mathrm{B}]_\mathrm{o}} R_i \sum\left(\frac{\partial[\mathrm{BL}]}{\partial K^{-1}}\right)^2 + \sum \frac{\partial[\mathrm{BL}]}{\partial[\mathrm{B}]_\mathrm{o}} \frac{\partial[\mathrm{BL}]}{\partial K^{-1}} \sum \frac{\partial[\mathrm{BL}]}{\partial K^{-1}} R_i}{\sum\left(\frac{\partial[\mathrm{BL}]}{\partial[\mathrm{B}]_\mathrm{o}}\right)^2 \sum\left(\frac{\partial[\mathrm{BL}]}{\partial K^{-1}}\right)^2 - \left(\sum \frac{\partial[\mathrm{BL}]}{\partial K^{-1}} \frac{\partial[\mathrm{BL}]}{\partial[\mathrm{B}]_\mathrm{o}}\right)^2},$$

$$\Delta K^{-1} = \frac{-\sum\left(\frac{\partial[\mathrm{BL}]}{\partial[\mathrm{B}]_\mathrm{o}}\right)^2 \sum \frac{\partial[\mathrm{BL}]}{\partial K^{-1}} R_i + \sum \frac{\partial[\mathrm{BL}]}{\partial[\mathrm{B}]_\mathrm{o}} R_i \sum \frac{\partial[\mathrm{BL}]}{\partial K^{-1}} \frac{\partial[\mathrm{BL}]}{\partial[\mathrm{B}]_\mathrm{o}}}{\sum\left(\frac{\partial[\mathrm{BL}]}{\partial[\mathrm{B}]_\mathrm{o}}\right)^2 \sum\left(\frac{\partial[\mathrm{BL}]}{\partial K^{-1}}\right)^2 - \left(\sum \frac{\partial[\mathrm{BL}]}{\partial K^{-1}} \frac{\partial[\mathrm{BL}]}{\partial[\mathrm{B}]_\mathrm{o}}\right)^2}.$$

When more than two parameters are to be fitted to a function, as for instance in the case of binding curves to non-identical sites, correspondingly larger matrices have to be evaluated. In the author's laboratory the Gauss–Jordan method has been found efficient for this purpose when small computers are used. The principle of this procedure for solving simultaneous equations relies on conversion of the matrix of coefficients in equation (A.8) into an identity matrix by suitable multiplications and subtraction of lines and compensating operations on the vector on the right-hand side of the equation (see for instance McCalla, 1967, p. 171).

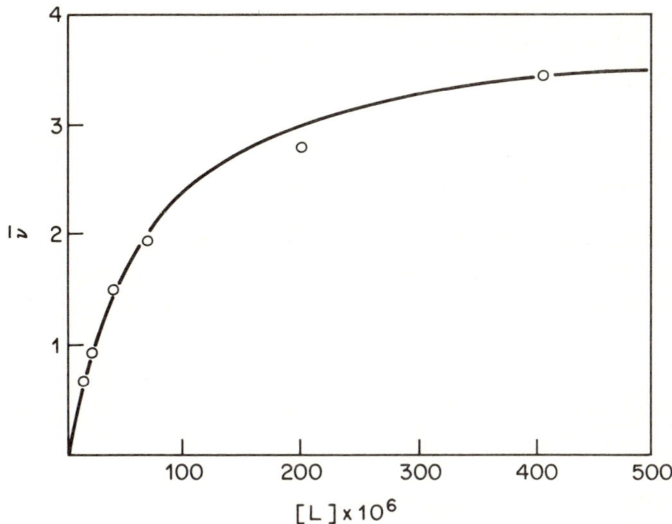

Figure A.2 Theoretical binding curve generated from least squares parameters and data points used for calculation in text

As an example one can use the data of Figure A.1 and consider $\bar{\nu}$ equal to the concentration $[BL] \times 10^6$ and $\bar{\nu}$ is equal to $[B]_o \times 10^6$ as [L] tends to infinity. If we take as a first estimate $K = 1.0 \times 10^4\ \text{M}^{-1}$ and $[B]_o = 3 \times 10^{-6}\ \text{M}$ we obtain

$(\Delta K + K) \times 10^{-4}$	$(\Delta[B]_0 + [B]_o) \times 10^6$	
1.000	3.000	Estimate
2.810	3.837	First correction
1.625	3.823	Second correction
1.483	3.886	Third correction
1.469	3.895	Fourth correction
1.468	3.896	Fifth correction

It is always of the utmost importance to plot the original raw data (experimental points) over the theoretical curve generated with the calculated parameters. Figure A.2 shows the binding curve generated from equation (A.7) with the parameters evaluated by the differential correction technique. The experimental points are also shown and it is clear that there are no systematic deviations. If in doubt one should plot the residuals, deviation of data points from the least square curve, to ascertain whether there are any non-random deviations. In this way one can detect non-random errors in the experimental technique and the possibility that the theoretical equation does not describe the behaviour of the system correctly.

In the present case the binding parameters obtained from the data using regression of the linearized equation and by the non-linear least squares technique give closely similar answers. Colquhoun (1971, p. 257) gives an example where the answers obtained from the two methods are markedly different. Only the parameters obtained from the non-linear squares technique results in a curve which fits the data without systematic deviations.

References

Ackers, G. K. (1980). *Biophys. J.*, **32**, 331.
Adair, G. S. (1923). *Science,* **58**, 13.
Adair, G. S. (1925a). *J. Biol. Chem.,* **63**, 529.
Adair, G. S. (1925b). *Proc. Roy. Soc. London A,* **109**, 292.
Adair, G. S. (1928). *Proc. Roy. Soc. London A,* **120**, 573.
Adelstein, R. S., and Eisenberg, E. (1980). *Ann. Rev. Biochem.,* **49**, 921.
Alexander, D. M., Hill, D. J. T., and White, L. R. (1971). *Aust. J. Chem.,* **24**, 1143.
Angell, C. A. (1983). In *Water: A Comprehensive Treatise*, Vol. 7 (F. Franks, ed.), Plenum Press, New York.
Angell, C. A., Shuppert, J., and Tucker, J. D. (1973). *J. Phys. Chem.,* **77**, 3092.
Armstrong, G. T. (1964). *J. Chem. Educ.,* **41**, 164.
Atkins, P. W. (1978). *Physical Chemistry*, Oxford University Press, London.
Baldwin, J. M., and Chothia, C. (1979). *J. Mol. Biol.*, **129**, 175.
Bates, R. G. (1973). *Determination of pH; Theory and Practice*, 2nd edn, Wiley, New York.
Beezer, A. E. (1980). *Biological Calorimetry*, Academic Press, London and New York.
Bell, J. E., and Dalziel, K. (1975). *Biochim. Biophys. Acta,* **391**, 249.
Benesch, R. E., and Benesch, R. (1955). *J. Amer. Chem. Soc.,* **77**, 5877.
Ben-Naim, A. (1974). *Water and Aqueous Solutions: Introduction to a Molecular Theory*, Plenum Press, New York.
Ben-Naim, A. (1978). *J. Phys. Chem.,* **82**, 792.
Ben-Naim, A. (1980). *Hydrophobic Interactions*, Plenum Press, New York.
Bernhard, S. A. (1956). *J. biol. Chem.,* **218**, 961.
Bjerrum, N. (1923). *Zeit. Phys. Chem.,* **104**, 406.
Bloomfield, V. A., Crothers, D. M., and Tinoco, I. (1974). *Physical Chemistry of Nucleic Acids*, Harper and Row, New York.
Bohr, C., Hassselbalch, K. A., and Krogh, A. (1904). *Skand. Arch. Physiol.* **16**, 401.
Borsook, H. (1953). *Adv. Protein Chem.,* **8**, 127.
Borsook, H., and Deasy, C. L. (1951). *Ann. Rev. Biochem.,* **20**, 209.
Borsook, H., and Schott, H. F. (1931). *J. Biol. Chem.,* **92**, 535.
Boyer, P.D., Chance, B., Ernster, L., Mitchell, P., Racker, E., and Slater, E. C. (1977). *Ann. Rev. Biochem.*, **46**, 955.
Brandts, J. F. (1965). *J. Amer. Chem. Soc.,* **87**, 2759.
Braunstein, J. (1969). *J. Chem. Educ.,* **46**, 719.
Bridgman, P. W. (1931). *Dimensional Analysis*, 2nd edn, Yale University Press, New Haven, Conn.
Brönsted, J. N., and LaMer, V. K. (1924). *J. Amer. Chem. Soc.,* **46**, 555.
Brown, W. E., and Hill, A. V. (1923). *Proc. Roy. Soc. London B,* **94**, 297.
Bruun, S. G., and Hvidt, A. (1977). *Ber. Bunsenges. Phys. Chem.,* **81**, 930.
Burton, K. (1974). *Biochem. J.,* **143**, 365.
Bulmer, M. G. (1979). *Principles of Statistics,* Dover, New York.
Butler, J. A. V. (1937). *Trans. Faraday Soc.,* **33**, 229.
Butler, J. A. V., Thompson, D. W., and McLennan, W. H. (1933). *J. Chem. Soc.,* **1933**, 674.
Buzzell, A., and Sturtevant, J. M. (1951). *J. Amer. Chem. Soc.*, **73**, 2454.

Cardwell, D. S. L. (1971). *From Watt to Clausius: The Rise of Thermodynamics in the Early Industrial Age*, Cornell University Press, Ithaca, N.Y.

Carnot, S. (1824). *Refexious sur la puissance motrice du feu*. Reprinted in 1978 by Librairie Philosophique, J. Vrin, Paris.

Cattell, McK. (1936). *Biol. Revs,* **11**, 441.

Chapman, D. (1975). *Quart. Rev. Biophys.,* **8**, 185.

Cherry, R. J. (1979). *Biochim. Biophys. Acta,* **559**, 289.

Christiansen, J., Douglas, C. G., and Haldane, J. S. (1914). *J. Physiol.,* **48**, 244.

Clark, W. M. (1960). *Oxidation–Reduction Potentials of Organic Systems*, Williams and Wilkins, Baltimore.

Coates, J. H., Hardman, M. J., Shore, J. S., and Gutfreund, H. (1977). *FEBS Letters,* **84**, 25.

Cohn, E. J., and Edsall, J. T. (1943). *Proteins, Amino Acids and Peptides,* Reinhold, New York. Reprinted in 1965 by Hafner, New York.

Colinvaux, P. (1980). *Why Big Fierce Animals are Rare*, Penguin Books, Harmondsworth.

Colosimo, A., Brunori, M., and Wyman, J. (1976). *J. Mol. Biol.,* **100**, 47.

Colquhoun, D. (1971). *Lectures on Biostatistics,* Oxford University Press, London.

Cornish-Bowden, A. (1976). *Principles of Enzyme Kinetics*, Butterworth, London.

Cornish-Bowden, A. (1980). *Fundamentals of Enzyme Kinetics,* Butterworth, London.

Dahlquist, F. W. (1978). *Methods Enzymol.,* **48**, 270.

Debye, P. (1927). *Zeit. Phys. Chem.,* Cohen Festband, 56. Reprinted in 1954 in *The Collected Papers of Peter J. W. Debye*, Interscience, New York, p. 336.

Debye, P., and Hückel, E. (1923). *Phys. Zeit.*, **24**, 185.

Denbigh, K. (1981). *The Principles of Chemical Equilibrium*, 4th edn, Cambridge University Press, Cambridge.

Dixon, M., and Webb, E. C. (1964). *Enzymes*, 2nd edn, Academic Press, New York and Longmans, London, p. 69.

Dobry, A., Fruton, J. S., and Sturtevant (1952). *J. Biol. Chem.,* **195**, 149.

Donnan, F. G. (1911). *Zeit. Electrochem.,* **17**, 572.

Dowd, J. E., and Riggs, D. S. (1965). *J. Biol. Chem.,* **240**, 863.

Eadie, G. S. (1942). *J. Biol. Chem.,* **146**, 85.

Edsall, J. T. (1953). In *The Proteins*, 1st edn, Vol. 1B (H. Neurath and K. Bailey, eds), Academic Press, New York, p. 549.

Edsall, J. T. (1969). In *CO_2: Chemical, Biochemical and Physiological Aspects* (R. E. Forster, ed.), NASA SP–188, US Government Printing Office, Washington, D.C., p. 15.

Edsall, J. T. (1974). *Mol. Cell. Biochem.,* **5**, 103.

Edsall, J. T., and McKenzie, H. A. (1978). *Adv. Biophys.,* **10**, 137.

Edsall, J. T., and Wyman, J. (1958). *Biophysical Chemistry*, Vol. I, Academic Press, New York and London. (The projected Volume II never appeared.)

Edsall, J. T., Felsenfeld, G., Goodman, D. S., and Gurd, F. R. N. (1954). *J. Amer. Chem. Soc.,* **76**, 3054.

Ehrenberg, M., and Rigler, R. (1976). *Quart. Rev. Biophys.,* **9**, 69.

Ellenbogen, E. (1952). *J. Amer. Chem. Soc.,* **74**, 5198.

Elwell, M. L., and Schellman, J. A. (1977). *Biochim. Biophys. Acta,* **494**, 387.

Engelborghs, Y., Heremans, K. A. H., De Mayer, L. C., and Hoebeke, J. (1976). *Nature,* **259**, 686.

Fersht, A. R. (1977). *Enzyme Structure and Function*, Freeman, Reading and San Francisco.

Fick, A. (1893). *Pflügers Arch. Ges. Physiol.,* **53**, 606.

Flint, H. T. (1958). *Nature,* **181**, 1098.

Flory, P. J. (1953). *Principles of Polymer Chemistry*, Cornell University Press, Ithaca, N.Y., Chapter XI.

Flory, P. J. (1969). *Statistical Mechanics of Chain Molecules*, Wiley-Interscience, New York.

Förster, Th. (1959). *Discuss. Faraday Soc.*, **27**, 7.

Franks, F. (ed.) (1972–82). *Water: A Comprehensive Treatise*, in seven volumes, Plenum Press, New York.

Franks, F. (1975). In *Water: A Comprehensive Treatise*, Vol. 4 (F. Franks, ed.), Plenum Press, New York, Chapter 1.

Franks, F., and Reid, D. S. (1973). In *Water: A Comprehensive Treatise*, Vol. 2 (F. Franks, ed.), Plenum Press, New York, Chapter 5.

Friend, S. H., and Gurd, F. R. N. (1979). *Biochemistry,* **18**, 4612 and 4620.

Garby, L., and Meldon, J. (1977). *The Respiratory Function of Blood*, Plenum, New York.

Gellert, M. (1981). *Ann. Rev. Biochem.*, **50**, 879.

Georgescu-Roegan, N. (1971). *The Entropy Law and the Economic Progress*, Harvard University Press, Cambridge, Mass.

Gerhart, J. C. (1970). *Curr. Top. Cell. Regul.,* **2**, 275.

German, B., and Wyman, J. (1937). *J. Biol. Chem.,* **117**, 533.

Gibbs, J. W. (1876). *Trans. Connecticut Acad. Sci.,* **III**, 343. Reprinted in 1931 in *The Collected Works of J. Willard Gibbs*, Longmans Green, London.

Gill, S. J., and Wadsö, I. (1976). *Proc. Natl Acad. Sci., U.S.A.,* **73**, 2955.

Gill, S. J., Nichols, N., and Wadsö, I. (1976). *J. Chem. Thermodynamics,* **8**, 445.

Gill, S. J., Gand, H. T., Wyman, J., and Barisas, B. G. (1978). *Biophys. Chem.,* **8**, 53.

Glaser, L., and Brown, D. H. (1955). *J. Biol. Chem.,* **216**, 67.

Goldberg, R. N., Prosen, E. J., Staples, B. R., Boyd, R. N., and Armstrong, G. T. (1975). *Anal. Biochem.,* **64**, 68.

Green, A. A. (1931). *J. Biol. Chem.,* **93**, 495 and 517.

Green, A. A. (1932). *J. Biol. Chem.,* **95**, 47.

Gross, P., and Iser, M. (1930). *Monatsh. Chem.,* **55**, 329.

Gross, P., and Schwartz, K. (1930). *Monatsh. Chem.,* **55**, 287.

Guggenheim, E. A. (1967). *Thermodynamics,* 5th edn, North Holland, Amsterdam.

Gutfreund, H. (1975). *Enzyme Physical Principles*, revised edn, Wiley-Interscience, Chichester and New York.

Gutfreund, H., and Trentham, D. R. (1975). *Ciba Foundation Symp.,* **31**, 69.

Haldane, J. B. S. (1930). *Enzymes*, Longmans, Green and Co., London. Reprinted in 1965 as an MIT Paperback by MIT Press, Cambridge, Mass.

Haldane, J. B. S., and Stern, K. G. (1932). *Allgemeine Chemie der Enzyme*, Dresden and Leipzig.

Harned, H. S., and Owen, B. B. (1958). *The Physical Chemistry of Electrolytic Solutions*, 3rd edn, Reinhold, New York.

Helmholtz, H. von (1847). *On the Conservation of Energy* (in German), G. Reimer, Berlin.

Henderson, L. J. (1913). *The Fitness of the Environment*, Macmillan, New York. Reprinted in 1958 by Beacon Press, Boston.

Hensley, P., Yang, Y. R., and Schachman, H. K. (1981). *J. Mol. Biol.*, **152,** 131.

Herriott, R. M., Desreux, V., and Northrop, J. H. (1940). *J. Gen. Physiol.* **24,** 213.

Henis, Y. I., and Levtizky, A. (1980). *Proc. Natl Acad. Sci. U.S.A.,* **77**, 5055.

Hill, A. V. (1910). *J. Physiol.,* **40**, IV.

Hill, A. V. (1913). *Biochem. J.,* **7**, 471.

Hill, A. V. (1965). *Trails and Trials in Physiology*, Arnold, London.

Holbrook, J. J. (1972). *Biochemistry,* **128**, 921.

Homsher, E., and Keen, C. J. (1978). *Ann. Rev. Physiol.,* **40**, 93.

Howarth, C. V. (1970). *Quart. Rev. Biophys.,* **3**, 429.

Hvidt, A. (1978). *Acta Chem. Scand. A,* **32**, 675.
Ihde, A. J., and Janssen, J. F. (1974). *J. Mol. Cell. Biochem.,* **5**, 11.
Imai, K., and Adair, G. S. (1977). *Biochim. Biophys. Acta,* **490**, 456.
Jacobson, G. R., and Stark, G. R. (1973). In *The Enzymes*, 3rd edn, Vol. 9 (P. D. Boyer, ed.), Academic Press, New York, p. 225.
Jencks, W. P. (1975). *Adv. Enzymol.,* **43**, 219.
Jencks, W. P. (1980). *Adv. Enzymol.,* **51**, 75.
Johnson, R. E., and Biltonen, R. L. (1976). *J. Amer. Chem. Soc.,* **97**, 2349.
Joule, J. P. (1849). *Phil. Trans. Roy. Soc. London,* **140**, 61.
Katchalsky, A., and Curran, P. F. (1965). *Nonequilibrium Thermodynamics in Biophysics*, Harvard University Press, Cambridge, Mass.
Katchalsky, A., and Spangler, R. (1968). *Quart. Rev. Biophys.,* **1**, 127.
Katz, B. (1966). *Nerve, Muscle and Synapse*, McGraw-Hill, New York.
Kauzmann, W. (1959). *Adv. Protein Chem.,* **14**, 1.
Kauzmann, W. (1967). *Thermodynamics and Statistics: with Applications to Gases*, Benjamin, New York.
Kauzmann, W. (1974). *Proc. 4th Internat. Conf. High Pressure, Kyoto,* 619.
Kell, G. S. (1967). *J. Chem. Eng. Data,* **12**, 67.
Keynes, R. D., and Aidley, D. J. (1981). *Nerve and Muscle*, Cambridge University Press, Cambridge.
Kleiber, M. (1961). *The Fire of Life*, Wiley, New York.
Kliman, H. (1969). Ph.D. thesis, Princeton University.
Klotz, I. M., and Hunston, D. E. (1971). *Biochemistry,* **10**, 3065.
Klotz, I. M., and Rosenberg, R. M. (1972). *Chemical Thermodynamics: Basic Theory and Methods*, 3rd edn, W. A. Benjamin, Menlo Park, Calif.
Kodama, T., and Woledge, R. C. (1976). *J. Biol. Chem.,* **251**, 7499.
Kodama, T., and Woledge, R. C. (1979). *J. Biol. Chem.,* **254**, 6382.
Kodama, T., Watson, J. D., and Woledge, R. C. (1977). *J. Biol. Chem.,* **252**, 8085.
Koltun, W. L., Clark, R. E., Dexter, R. N., Katsoyannis, P. G., and Gurd, F. R. N. (1959). *J. Amer. Chem. Soc.,* **81**, 295.
Konicek, J., and Wadsö, I. (1971). *Acta Chem. Scand.,* **25**, 1541.
Konigsberger, L. (1965). *Hermann von Helmholtz* (translated by F. A. Welby), Dover, New York. (Originally published in German in 1902.)
Krebs, H. A. (1953). *Biochem. J.,* **54**, 78.
Krishnan, K. S., and Brandts, J. F. (1978). *Methods Enzymol.,* **49**, 3.
Kuffler, S. W., and Nichols, J. G. (1976). *From Neuron to Brain*, Sinauer Associates, Sunderland, Mass.
Ladbrooke, B. D., and Chapman, D. (1969). *Chem. Phys. Lipids,* **3**, 304.
Landsborough, J. C., and Dalziel, K. (1968). *Biochem. J.,* **110**, 217.
Langerman, N., and Biltonen, R. L. (1979). *Methods Enzymol.,* **61**, 261.
Le Chatelier, H. L. (1888). *Ann. mines,* **8**, 157; see also *Zeit. Phys. Chem.,* **9**, 335 (1892).
Lehmann, J. (1930). *Skand. Arch. Physiol.,* **58**, 173.
Lehninger, A. L. (1975). *Biochemistry,* 2nd edn, Worth, New York.
Lewis, G. N., and Randall, M. (1923). *Thermodynamics*, McGraw-Hill, New York.
Lewis, G. N., and Randall, M. (1961). *Chemical Thermodynamics*, 2nd edn (revised by K. S. Pitzer and L. Brewer), McGraw-Hill, New York.
Linden, C. D., and Fox, C. F. (1975). *Acc. Chem. Res.,* **8**, 321.
Lipmann, F. (1941). *Adv. Enzymol.,* **1**, 99.
Long, F. A., and McDevit, W. F. (1952). *Chem. Rev.,* **51**, 119.
Mabrey, S. and Sturtevant, J. M. (1976). *Proc. Natl Acad. Sci. U.S.A.,* **73**, 3862.
McCalla, T. R. (1967). *Introduction to Numerical Methods and Fortran Programming*, Wiley, New York.
McCubbin, W. D., and Kay, C. M. (1973). *Biochemistry,* **12**, 4228.

McGhee, J. D., and von Hippel, P. H. (1974). *J. Mol. Biol.,* **86**, 469.
McGlashan, M. L. (1979). *Chemical Thermodynamics*, Academic Press, London.
Mannherz, H. G., Schenck, H., and Goody, R. S. (1974). *Europ. J. Biochem.,* **48**, 287.
Margenau, H., and Murphy, G. M. (1943). *The Mathematics of Physics and Chemistry*, Van Nostrand, New York.
Martin, R. B., Edsall, J. T., Wetlaufer, D. B., and Hollingworth, B. R. (1958). *J. Biol. Chem.,* **233**, 1429.
Masterton, W. L. (1954). *J. Chem. Phys.,* **22**, 1830.
Mateo, P. L., and Privalov, P. L. (1981). *FEBS Letters,* **123**, 189.
Matthew, J. B., Hanania, G. I. H., and Gurd, F. R. N. (1979). *Biochemistry,* **18**, 1919.
Maxwell, J. C. (1872). *Theory of Heat*, 3rd edn, Longman, London.
Meyerhof, O. (1930). *Die chemischen Vorgänge im Muskel*, Springer, Berlin.
Meyerhof, O., and Schulz, W. (1935). *Biochem. Z.,* **281**, 292.
Michaelis, L. (1940). *Ann. N.Y. Acad. Sci.,* **40**, 39.
Mills, F. C., and Ackers, G. K. (1979). *J. Biol. Chem.,* **254**, 2881.
Moore, W. J. (1963). *Physical Chemistry*, 4th edn, Prentice-Hall, Englewood Cliffs, N. J., and Longman Green, London.
Mountcastle, D. B., Biltonen, R. L., and Halsey, M. J. (1978). *Proc. Natl Acad. Sci. U.S.A.,* **75**, 4906.
Needham, D. M. (1971). *Machina Carnis: The Biochemistry of Muscle Contraction in its Historical Development*, Cambridge University Press, Cambridge.
Nernst, W. (1906). *Nachr. Kgl. Ges. Wiss., Göttingen, Math.-Phys. Kl.,* **1906**, 1.
Newsholme, E. A., and Start, C. (1973). *Regulation in Metabolism*, Wiley, Chichester and New York.
Nichols, N., Sköld, R., Spink, O., Suurkursk, J., and Wadsö., I. (1976). *J. Chem. Thermodynamics,* **8**, 1081.
Norby, J. G., Ottolenghi, P., and Jensen, J. (1980). *Anal. Biochem.,* **102**, 318.
Nozaki, Y., and Tanford, C. (1967). *J. Biol. Chem.,* **242**, 4731.
Nozaki, Y., Gurd, F. R. N., Chen, R., and Edsall, J. T. (1957). *J. Amer. Chem. Soc.,* **79**, 2123.
Perron, G., and Desnoyers, J. E. (1979). *Fluid Phase Equilibria,* **2**, 239.
Perutz, M. F. (1980). *Proc. Roy. Soc. London B,* **208**, 135.
Phillips, R. C., George, P., and Rutman, R. J. (1963). *Biochemistry,* **2**, 501.
Planck, M. (1911). *Thermodynamik,* 3rd edn, Veight & Co., Leipzig.
Popper, K. (1976). *Unended Quest*, Fontana-Collins, London.
Privalov, P. L. (1974). *FEBS Letters,* **40**, S140.
Privalov, P. L. (1979) *Adv. Protein Chem.*, **33**, 167; (1982) **35**, 1.
Radda, G. K. (1975). *Phil. Trans. Roy. Soc. London B,* **272**, 171.
Rao, B. D. N., Buttlaire, D. H., and Cohn, M. (1976). *J. Biol. Chem.,* **251**, 6981.
Rao, B. D. H., Cohn, M., and Noda, L. (1978). *J. Biol. Chem.,* **253**, 1149.
Reynolds, J. A., Gilbert, D. B., and Tanford, C. (1974). *Proc. Natl Acad. Sci. U.S.A.,* **71**, 2925.
Rigler, R., and Ehrenberg, M. (1976). *Quart. Rev. Biophys.,* **9**, 1.
Robbins, E. A., and Boyer, P. D. (1957). *J. Biol. Chem.*, **224**, 121.
Roberts, D. V. (1977). *Enzyme Kinetics*, Cambridge University Press, Cambridge.
Rosing, J., and Slater, E. C. (1972). *Biochim. Biophys. Acta,* **267**, 275.
Robinson, R. A., and Stokes, R. H. (1970). *Electrolyte Solutions*, 2nd edn, Butterworth, London.
Rossi Fanelli, A., Antonini, E., and Caputo, A. (1964). *Adv. Protein Chem.,* **19**, 73.
Rushbrooke, G. S. (1949). *Introduction to Statistical Mechanics*, Oxford University Press, Oxford.
Scatchard, G. (1946). *J. Amer. Chem. Soc.,* **68**, 2315 and 2320.
Scatchard, G. (1949). *Ann. N.Y. Acad. Sci.,* **51**, 660.

Scatchard, G. (1976). *Equilibrium in Solutions*, Harvard University Press, Cambridge, Mass.

Scatchard, G., Batchelder, A. C., and Brown, A. (1944). *J. Clin. Invest.*, **23**, 458.

Scatchard, G., Coleman, J. S., and Shen, A. L., (1957). *J. Amer. Chem. Soc.*, **79**, 12.

Scatchard, G., Scheinberg, T. H., and Armstrong, S. H. (1950). *J. Amer. Chem. Soc.*, **72**, 535 and 540.

Scatchard, G., Wu, Y.V., and Shen, A. L. (1959). *J. Amer. Chem. Soc.*, **81**, 6104.

Schachman, H. K. (1959). *Ultracentrifugation in Biochemistry*, Academic Press, New York.

Schmidt, D. E., and Westheimer, F. H. (1971). *Biochemistry*, **10**, 1249.

Schreier, A. A., and Schimmel, P. R. (1974). *J. Mol. Biol.*, **86**, 601.

Shannon, C. E., and Weaver, W. (1949). *The Mathematical Theory of Communication*, University of Illinois Press, Urbana, Ill.

Smith, P. K., and Smith, E. R. B. (1942). *J. Biol. Chem.*, **146**, 187.

Smith, R., and Tanford, C. (1973). *Proc. Natl Acad. Sci. U.S.A.*, **70**, 289.

Snow, C. P. (1959). *The Two Cultures*, Cambridge University Press, Cambridge.

Sörensen, S. P. L. (1909). *Biochem. Z.*, **21**, 131.

Stevens, C. F. (1977). *Nature*, **270**, 391.

Sturtevant, J. M. (1955). *J. Amer. Chem. Soc.*, **77**, 255.

Sturtevant, J. M. (1974). *Ann. Rev. Biophys.*, **3**, 35.

Sturtevant, J. M. (1976). *Proc. Natl Acad. Sci. U.S.A.*, **73**, 3862.

Sturtevant, J. M. (1977). *Proc. Natl Acad. Sci. U.S.A.*, **74**, 2235.

Svedberg, T., and Pedersen, K. O. (1940). *The Ultracentrifuge*, Oxford University Press, Oxford.

Szilard, L. (1925). *Zeit. Phys.*, **53**, 753.

Tanford, C. (1961). *Physical Chemistry of Macromolecules*, Wiley, New York.

Tanford, C. (1962). *Adv. Protein Chem.*, **17**, 69.

Tanford, C. (1980). *The Hydrophobic Effect*, 2nd edn, Wiley-Interscience, New York.

Tiktopulo, E. I., and Privalov, P. L. (1978). *FEBS Letters*, **91**, 57.

Traube, J. (1899). *Sammlung Chem. Chem. Tech. Vorträge*, **4**, 255.

Treloar, L. R. G. (1958). *The Physics of Rubber Elasticity*, Oxford University Press, Oxford.

Trentham, D. R., Eccleston, J. F., and Bagshaw, C. (1976). *Quart. Rev. Biophys.*, **9**, 2.

Tyuma, I., Imai, K., and Schimizu, K. (1972). In *Oxygen Affinity of Haemoglobin and Red Cell Acid–Base Status*, Fourth Alfred Benzon Symposium, Munksgaard, Copenhagen, and Academic Press, New York, p. 131.

Van't Hoff, J. H. (1886). *Arch. Néerl. Sci.* **20**, 239.

Veech, R. L., Raijam, L., Dalziel, K., and Krebs, H. A. (1969). *Biochem. J.*, **115**, 837.

Veech, R. L., Lawson, J. W. R., Cornell, N. W., and Krebs, H. A. (1979). *J. Biol. Chem.*, **254**, 6538.

Villet, R. H., and Dalziel, K. (1969). *Biochem. J.*, **115**, 633.

Wadsö, I. (1970). *Quart. Rev. Biophys.*, **3**, 383.

Wagner, R. H. (1949). In *Physical Methods of Organic Chemistry*, 2nd edn, part I, Interscience, New York, p. 487.

Walz, D. (1979). *Biochim. Biophys. Acta*, **505**, 279.

Weber, G. (1975). *Adv. Protein Chem.*, **29**, 1.

Whiffen, D. H. (1979). *Pure Appl. Chem.*, **51**, 1.

Whitaker, J. R., Yates, D. W., Bennett, N. G., Holbrook, J. J., and Gutfreund, H. (1974). *Biochem. J.*, **139**, 677.

Whitehead, E. P. (1978). *Biochem. J.*, **171**, 501.

Whitehead, E. P. (1980a). *J. Theoret. Biol.,* **86**, 45.
Whitehead, E. P. (1980b). *J. Theoret. Biol.,* **87**, 153.
Wilhelm, E., Battino, R., and Wilcock, R. J. (1977). *Chem. Rev.,* **77**, 219.
Williams, J. W. (ed.) (1963). *Ultra-centrifugal Analysis in Theory and Experiment*, Academic Press, New York and London.
Williams, J. W. (1972). *Ultracentrifugation of Macromolecules*, Academic Press, New York.
Williams, R. J. P. (1975). *Biochim. Biophys. Acta,* **416**, 237.
Wilson, S. S. (1981). *Scient. Amer.,* **245**, 134.
Wimmer, M. J., and Rose, I. A. (1978). *Ann. Rev. Biochem.,* **47**, 1031.
Winkler-Oswatitsch, R., and Eigen, M. (1979). *Angew. Chem. Int. Ed.,* **18**, 20.
Winlund, C. C., and Chamberlin, M. J. (1970). *Biochem. Biophys. Res. Comm.,* **40**, 43.
Woledge, R. C. (1977). *Applications of Calorimetry in Life Sciences*, Walter de Gruyter, Berlin and New York, p. 183.
Wyman, J. (1948). *Adv. Protein Chem.,* **4**, 407.
Wyman, J. (1964). *Adv. Protein Chem.,* **19**, 223.
Wyman, J. (1967). *Quart. Rev. Biophys.,* **1**, 35.
Wyman, J. (1975). *Proc. Natl Acad. Sci. U.S.A.* **72**, 1464.

Index